Biomass for Bioenergy - Recent Trends and Future Challenges

Edited by Abd El-Fatah Abomohra

Published in London, United Kingdom

IntechOpen

Supporting open minds since 2005

Biomass for Bioenergy - Recent Trends and Future Challenges
http://dx.doi.org/10.5772/intechopen.77410
Edited by Abd El-Fatah Abomohra

Contributors
Sakthivadivel D, Manikandan S, Kumaraselvan M, Ganeshkumar Poongavanam, Iniyan S, Hamid Rezaei, Fahimeh Yazdanpanah, Shahab Sokhansanj, C. Jim Lim, Anthony Lau, Muhammad Naiman Aftab, Irfana Iqbal, Meisam Tabatabaei, Fatima Riaz, Ahmet Karadag, Alissara Reungsang, Apilak Salakkam, Pensri Plangklang, Sureewan Sittijunda, Mallika Kongkeitkajorn, Siriporn Lunprom, Sibel Irmak, Bushra Sadia, Faisal Saeed Awan, Fozia Saleem, Ali Razzaq, Bushra Irshad, Azlan Shah Hussain, Najiah Nadir, Nur Liyana Ismail, Manuel Paneque, Celián Román-Figueroa, Sebastián Herrera, Elizabeth Quintana, Flora Itzel Beltran-Ramirez, Ivette Cornejo-Corona, Domancar Orona-Tamayo, Jose De Jesus Esparza-Claudio, Jose Luz Nicasio Gonzalez-Cervantes

First published in London, United Kingdom, 2019 by IntechOpen
IntechOpen is the global imprint of INTECHOPEN LIMITED, registered in England and Wales, registration number: 11086078, 7th floor, 10 Lower Thames Street, London, EC3R 6AF, United Kingdom
Printed in Croatia

British Library Cataloguing-in-Publication Data
A catalogue record for this book is available from the British Library

Additional hard and PDF copies can be obtained from orders@intechopen.com

Biomass for Bioenergy - Recent Trends and Future Challenges
Edited by Abd El-Fatah Abomohra
p. cm.
Print ISBN 978-1-78923-987-4
Online ISBN 978-1-78923-988-1
eBook (PDF) ISBN 978-1-83968-221-6

For EU product safety concerns:
IN TECH d.o.o., Prolaz Marije Krucifikse Kozulić 3, 51000 Rijeka, Croatia, info@intechopen.com or visit our website at intechopen.com.

Meet the editor

Abd El-Fatah Abomohra is a Professor at the Jiangsu University, China and an Associate Professor at the Tanta University, Egypt. During 18 years of teaching and research experience, he has successfully undertaken research programs for MSc, PhD, post-doctoral studies, and as a faculty member. His research focuses on using microorganisms for enhanced bioenergy production, waste-water treatment, and CO2 sequestration in addition to production of valuable compounds. He has supervised many MSc and PhD students in Egypt and China. In addition, he has participated in a number of national projects in Egypt, Germany, and China for large biomass and biofuel production from microalgae. He has taught many undergraduate and postgraduate courses during his career.

Contents

Preface

Global demand and environmental pollution are the two inevitable issues that dictate the urge for finding alternative energy sources. The energy demand is positively correlated to population growth, as population increases, this directly increases the sophistication and eventual energy demand. From another perspective, environmental concerns are alarming, such as the adversity of global warming because of fossil fuel usage. It is evident and reported that the continued use of fossil fuels has resulted in excessive emissions of greenhouse gases into the atmosphere, leading to the depletion of the ozone layer and this has led to global warming. To address both the issues of an unprecedented increase in energy demand and environmental adversity, bioenergy has become appropriately attractive because of its renewable and eco-friendly nature. For almost two decades, biofuel research has received more interest and eventual growth and development. In general, it is widely accepted that biomass-derived products can reduce the current dependence on fossil-based products.

Biomass is the most abundant photosynthetic renewable resource on Earth, with a huge global annual production rate. Lignocellulosic wastes have been widely discussed as a promising natural chemical source and alternative feedstock for second-generation biofuels, without competition on the human food demand, fresh water, or arable lands. However, there are still many technical and economic challenges facing its utilization. Lignin is one of the components of lignocellulosic biomass. It is the most rigid constituent and can be considered as a glue providing the cell wall with stiffness and the plant tissue with compressive strength. In addition, it provides resistance to chemical and physical damage. Resistance of lignocelluloses to hydrolysis is mainly from the protection of cellulose by lignin and cellulose binding to hemicellulose. The present book provides basic knowledge and recent research on different applications of biomass, focusing on the bioenergy and different pretreatment methods that overcome the aforementioned hurdles. The book is divided into two main sections with a total of eleven chapters. The first section focuses on the challenges of biomass utilization and the different methods used for the pretreatment of lignocelluloses. The second section presents experimental studies and a literature survey for recent applications of biomass in biofuel production.

I would like to express my gratitude to the faculty members of the School of Energy and Power Engineering and to the postgraduate students of the New Energy Department, School of Energy and Power Engineering, Jiangsu University. Last but not least, I would like to thank Mr. Ivan Butkovic, Author Service Manager, IntechOpen, for being generously helpful throughout the process of submission and review.

Abd El-Fatah Abomohra
New Energy Department,
School of Energy and Power Engineering,
Jiangsu University,
China

Section 1

Challenges in Biomass Utilization

Chapter 1

Challenges of Biomass Utilization for Biofuels

Sibel Irmak

Abstract

Lignocellulosic biomass materials are attracting increasing attention as renewable, economical and abundant resources to reduce dependency on petroleum resources. However, chemical and physicochemical properties of these materials (e.g., low density, moisture content, complex and rigid structure, etc.) limit their use. The contents and compositions of cellulose, hemicellulose and lignin polymers in biomass feedstock highly affect the efficiencies of conversion technologies. In aqueous-phase utilization processes, it is necessary to breakdown the starting biomass material into soluble components and release the carbohydrates into hydrolysate for an effective conversion. It is not economical to convert biomass hydrolysates into renewable fuels in high yields if they mostly contain large-sized molecules such as polysaccharides. The chapter describes challenges of lignocellulosic biomass materials usage in biofuel application such as their complexity and diversity, content and composition, low density, moisture content and dissolution difficulties, etc.

Keywords: biomass, utilization, conversion, biofuels, challenges

1. Introduction

Non-edible lignocellulosic biomass materials are attracting increasing attention as renewable, economical, and abundant resources to reduce dependency on petroleum resources and minimize energy and material feedstock costs. These resources do not cause additional increase in the carbon dioxide level in the earth's atmosphere compared to fossil-based energy fuels such as coal, gasoline or natural gas. The carbon dioxide captured in biomass growth mostly balances with the release of carbon dioxide from bioenergy/biofuel. Therefore, use of biomass energy has the potential to reduce greenhouse gas emissions.

Biomass materials are the largest carbon sources for production of various fuels, chemicals and platform compounds and bioproducts. Because of their heterogeneous, complex and rigid structures it is hard to breakdown these materials to smaller components and/or convert to a wide range of value-added products. Biomass has a relatively low energy density; therefore, it requires more biomass feedstocks to supply the same amount of energy as a traditional hydrocarbon fuel. High oxygen contents of biomass materials can also negatively affect their conversion to various products such as fuels. For instance, in order to produce hydrocarbons fuels that can be comparable with petroleum-based ones oxygen should be removed from biomass structure. Efficiency of conversion processes can also varied depending on the biomass types (hardwood, softwood, grass, etc.).

Following sections will focus on the challenges for conversion of biomass to biofuels in detail.

2. Challenges

2.1 Moisture content

Biomass materials with high moisture contents is not a suitable feedstock for conventional thermochemical conversion technologies such as gasification and pyrolysis. High moisture can reduce the effectiveness of conversion processes. Moisture in raw biomass materials is also undesired because fuel produced from these materials can contain moisture. The fuels, which have high moisture contents, cannot burn easily. Some part of energy in the fuel are consumed for vaporization of water, which is present in the fuel. In order to maximize the heating value of the fuel produced from these materials the moisture content of biomass should be less than 20%. Drying the materials before being used in the conversion process is not preferable because of high cost. On the other hand, some biomass conversion processes use biomass with high moisture contents. For instance, hydrothermal conversion processes, which use supercritical and subcritical water as reaction medium, and biological processes such as alcohol production from carbohydrates by biomass hydrolysis and fermentation can be applied to the biomass with high moisture content without the need for drying. In these processes, moisture in the biomass plays an important role in the conversion, either as a major reactant, or as a reaction environment.

High moisture content in biomass causes biological degradation, mold formation and losses in the organic contents (e.g., carbohydrates) during storage [1], that could reduce the yield of the fuel produced from these materials. Storing biomass at <10% can extend the conservation time of the materials and reduce major losses (sugars) in the biomass during the storage period [2]. The drawbacks of high moisture contents can be mostly solved by compressing the biomass material for more uniform properties that process is called densification. Increasing bulk density of biomass materials by densification reduces transportation costs and storage volume. However, this process increases the price of the end product.

2.2 Density

The bulk density of lignocellulosic biomass materials is generally low (80–150 kg/m^3 for grass biomass and 160–220 kg/m^3 for woody biomass). This creates difficulties to handle such large quantities of feedstocks and increases their transportation and storage costs [3]. The bulk density of biomass should be between 190 and 240 kg/m^3 for efficient transport in various sizes of trucks with 25 ton loads [4].

The size, shape, moisture content, particle density, and surface characteristics are the factors affecting the bulk density of a material. The challenge for low density and different size and shapes of biomass can be overcome by densification process (**Figure 1**). In this process, biomass materials are mechanically compressed to increase their density and convert them to uniform shapes and sizes (briquetting, pelletizing, or cubing) [5, 6].

The density of biomass material can be increased ten-fold depending to biomass type, moisture content, processing conditions, etc. The costs of handling, transportation, and storage of resulted densified materials can be considerably reduced. Because of uniform sizes and shapes, the materials can be easily handled with standard machines or equipment [6].

Figure 1.
Increasing density of biomass.

2.3 Complexity and diversity

Lignocellulosic biomass materials is mainly composed of three components which are lignin, cellulose, and hemicellulose (**Figure 2**). These polymers are organized in complex non-uniform three-dimensional structures and each one has different polymerization degrees. Polymerization degree and/or structures of these biopolymers can vary among biomass species. Cellulose is a linear structure composed of β(1–4) linked glucose subunits. Cellulose molecules determine the cell wall

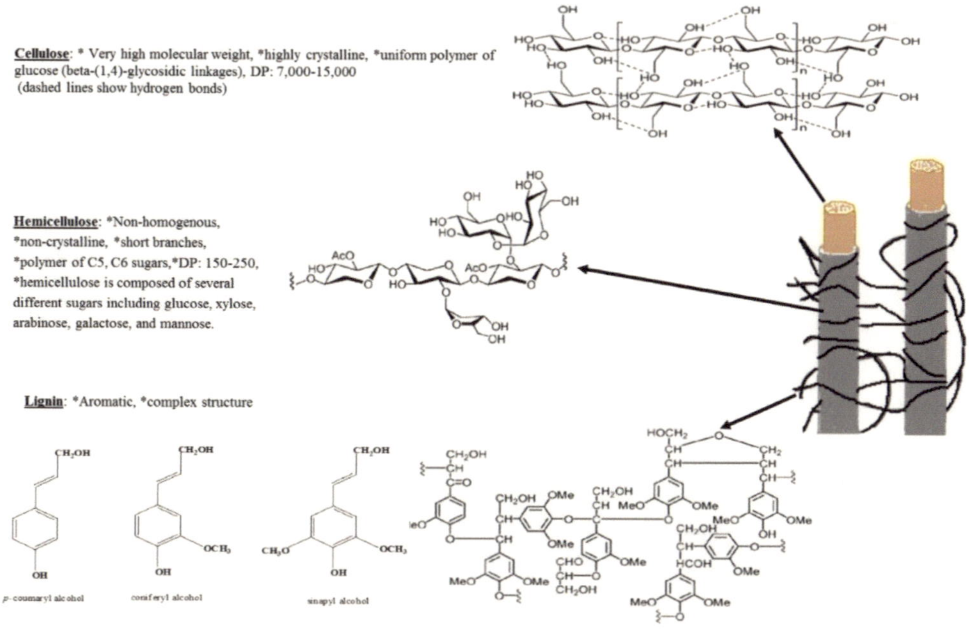

Figure 2.
Structure of lignocellulosic biomass.

framework. The inter- and intra- chain hydrogen bonding in the structure makes the cellulose to be crystalline and this portion of cellulose does not hydrolyze easily compared to amorphous cellulose structure [7, 8]. Hemicellulose has a random and amorphous structure, which is composed of several heteropolymers such as xylan, galactomannan, arabinoxylan, glucomannan and xyloglucan. Its polymerization degree is less than cellulose. The monomer units of hemicellulose polysaccharide include xylose, mannose, galactose, rhamnose, and arabinose units unlike only glucose in cellulose. Lignin is a complex aromatic substance of phenyl propane units. Three different phenyl propane building blocks p-coumaryl alcohol, coniferyl alcohol, and sinapyl alcohol, forms lignin structure.

Phenylpropanoid monomeric units in the lignin polymer are identified as p-hydroxyphenyl (H), guaiacyl (G), and syringyl (S) units, respectively. The ratio of these units varies between plants; for example in hardwoods S and G forms dominate with minor amount of p-hydroxyphenyl (H), whereas softwood lignins contain only G units. On the other hand, lignins from grasses are composed of the three basic precursors (HGS) [9, 10].

Composition of lignin, cellulose and hemicellulose in biomass materials significantly differ among biomass species (**Table 1**). For instance, some biomass materials such as hardwoods contain more cellulose in their structures, while others such as straws have more of hemicelluloses. Hemicellulose fractions of softwoods mainly have D-mannose derived structures such as galactoglucomannans, while hemicelluloses in hardwoods have D-xylose derived structures such as arabinoglucuronoxylan [13]. This diversity among biomass materials can significantly affect the conversion processes for production of biofuel or other useful products from biomass materials.

2.4 Recalcitrance and dissolution difficulties

Despite their potential for the production of biofuels and chemicals alternative to petroleum, the complex and rigid structures of lignocellulosic materials limit their use in such applications. Success of using lignocellulosic biomass for biofuels and other useful chemical productions depends largely upon physical and chemical properties of the biomass, on pretreatment methods and optimization of the processing conditions. The compositional changes in plant cell wall and differences in ultrastructure greatly influence the pretreatment and hydrolysis (dissolution) efficiency of the biomass. Hydrolysis is a chemical reaction that releases sugars from biomass structures. Biomass dissolution involves both physical, chemical and/

Biomass material	Cellulose	Hemicellulose	Lignin
Switchgrass (grass)	33.8	28.4	16.6
Miscanthus (grass)	47.7	24.6	12.3
Poplar (hardwood)	52.1	27.4	15.9
Oak (hardwood)	40.4	35.9	24.1
Pine (softwood)	46.0	25.5	20
Spruce (softwood)	45.5	22.9	27.9
Corn stover (agricultural waste)	38.5	24.5	18.5
Rice husks (agricultural waste)	32.1	20.6	17.7
Corn bran (byproduct of milling)	20.5	65.3	1.6

Table 1.
Various lignocellulosic biomass materials and their chemical compositions [11, 12].

or thermochemical treatment processes. The crystallinity of cellulose, hydrophobicity of lignin, and embedding the cellulose in lignin-hemicellulose matrix and difficulties in cleavage of some linkages (hydrogen bonding, ether linkages between the phenyl propane units, etc.) make biomass materials resistant to hydrolysis. It is relatively easy to hydrolyze hemicellulose into simple sugars compared to cellulose because hemicellulose fraction is more accessible compared to cellulose fraction which is embedded in the lignin matrix.

Biomass materials must first be broken down into components with smaller molecular weights (e.g., oligo- and monosaccharides) in order to be efficiently converted into a range of products. Hydrolysates from biomass can be used for producing a wide range of value-added products, including biofuels (ethanol, hydrogen, etc.), industrially important chemicals (e.g., solvents), and food products (sugar and sugar alcohols, etc.).

Significant existing challenges for hydrolysis of lignocellulosic biomass materials include the following:

- Existing hydrolysis methods are expensive and time consuming. Most of them are not environmentally friendly.

- Additional steps are required (pretreatment, neutralization, etc.)

- Released carbohydrates decompose in harsh hydrolysis conditions.

The major hydrolysis processes typically used for the solubilization of biomass require either use of toxic, corrosive, and hazardous chemicals (e.g., acid and alkali treatments) or longer retention times (e.g., enzymatic hydrolysis), which collectively make the process environmentally unsafe and/or expensive. Mineral acids are commonly used to dissolve hemicelluloses, whereas lignin is typically dissolved by alkaline or organosolv pretreatments [14, 15]. Recovery of the chemical catalyst is often crucial to the success of these processes [16]. On the other hand, generally harsh conditions (e.g., high temperatures and high acid concentrations) are needed to release glucose from biomass complex structures. Pyrolysis and other side reactions at higher temperatures become very important, and the amount of undesirable byproducts (tars) increases as the temperature is increased above 220°C [17].

Concentrated acid hydrolysis has been applied to breakdown lignocellulosic efficiently [18–20]. The hydrolysis of cellulose to its monomer sugar component occurs by degradation of chemical bonds in cellulose by the hydrolytic cleavage of β-1, 4-glycosidic bond which is catalyzed by H^+ ions of an acid. The reaction rate depends on the H_3O^+ ion concentration, the reaction temperature, and the chemical environment of the glycosidic bond and the rate is increased with the increasing acid ion concentration and temperature. The acid hydrolysis process usually employs sulfuric acid and hydrochloric acid at concentrations of 1–10% using a moderate temperature (in the range of 100–150°C) [21]. A two-step sulfuric acid hydrolysis is a widely used technique for releasing sugars from biomass [18]. Biomass is first treated with concentrated sulfuric acid at a low temperature and then hydrolyzed with diluted sulfuric acid at an elevated temperature. Concentrated acid recrystallizes cellulose to less crystallized oligosaccharides followed by less concentrated and higher reaction temperature for converting recrystallized oligosaccharides to monosaccharides. Concentrated acid hydrolysis process can provide higher conversion from polysaccharides to monosaccharides with minimum formation of reaction by-products with careful control of reaction conditions.

The use of concentrated acid for biomass hydrolysis has several more drawbacks such as energy consumption, equipment corrosion, handling of non-safe chemicals, an added necessary step of acid neutralization, the formation of byproducts that create an inhibitory effect in the fermentation [22, 23] and other negative environmental impacts. Thus, the current methods have undesirable processes and do not meet the needs.

Subcritical water (99.97°C < T < 374.15°C; 217.76 atm < P) is an alternative way to hydrolyze lignocellulosic biomass in an environmentally friendly manner by only operating temperature and pressure conditions. Significant advantages of subcritical water over commonly used biomass breakdown methods—alkali, acidic, and enzymatic—are summarized in **Table 2**.

The chemical properties of water are greatly changed at high temperatures and pressures due to the reduction of hydrogen bonding, which causes changes in dissociation, solubility, diffusivity, and reactivity [24]. Subcritical water has a lower relative dielectric constant and a higher ionic product than ambient water. When the temperature of water increases from ambient temperature to 250°C, its relative dielectric constant decreases from around 80 to nearly 27, which is similar to that of acetone at ambient temperature [25, 26]. Furthermore, the ion product of subcritical water substantially increases with temperature; therefore, subcritical water can catalyze chemical reactions such as hydrolysis and degradation without the use of any additional catalyst [27, 28].

Ionic product numbers of water (Kw) at various temperatures and pressures showed that when pressure is around 35 MPa and temperature is in sub- and supercritical regions under 400°C, Kw values are always higher than 1×10^{-14}. The Kw increases to its maxima (~10^{-11}) between 200 and 300°C and does not respond to changes in pressure when in this temperature range. The molar concentrations of hydrogen ion (H^+) and hydroxide ion (OH^-) in these regions are almost 30 times higher than those under room temperature. Therefore, the hydrolysis yield in these regions is expected to be high, and biomass polymers could be broken down into their smaller molecular weight components efficiently [29–31].

The presence of a weak acid in subcritical water media can also improve hydrolysis of biomass materials. The use of carbon dioxide is as a pressurizing gas caused the formation of carbonic acid that plays a catalytic role in effective solubilization of biomass [32]. Some studies indicated that the addition of small amounts of hydrogen peroxide can enhance lignin removal and modify cellulose structure toward favoring enzymatic hydrolysis [33, 34].

Complexity and diversity of the biomass materials considerable affect the solubilization efficiency of these materials. The differences in the content and composition of resulted hydrolysates can change the yield of the biofuel or target compound produced from these biomass hydrolysates. The more degraded organics containing hydrolysates can positively affect the yield of certain various value-added products; for instance, production of gaseous products by hydrothermal gasification

Breakdown methods	Advantages and disadvantages
Alkali/acidic	Requires harsh conditions; uses corrosive, hazardous chemicals; high costs of chemicals; formation of inhibitory byproducts; recovery problem of chemicals; requires a neutralization step; decomposes released sugars.
Enzymatic	Ineffective unless coupled with an acid treatment; high cost; time consuming.
Subcritical water	Takes place in a safe solvent (water); no corrosive, hazardous, or toxic chemicals are needed; hydrolysis efficiency can be enhanced by operating temperature and pressure; use of recyclable heterogeneous catalysts can make the process more effective.

Table 2.
Comparison of alkali, acidic and enzymatic biomass breakdown methods with subcritical water treatment.

processes. The maximum solubilization yield of wheat straw and kenaf biomass materials was 70–75%, which was achieved at 250°C in subcritical water medium [32, 35]. However, the hydrolysates obtained in this process had high molecular weight polysaccharides that were difficult to utilize for hydrogen production by aqueous-phase reforming [31]. For maximum usability, biomass components in hydrolysates should be further broken down into smaller molecular weight components with a suitable method.

2.5 Other challenges

Although energy demands are continuous, biomass materials are seasonal. Some biomass feedstocks have advantages in terms of production, harvesting, storage, and transportation compared to others. Non-food biomass such as energy crops (switchgrass, miscanthus, kenaf, etc.) have advantages over food crops (corn, sugarcane, sugar beet, sweet sorghum, etc.). Perrenial energy crops such as switchgrass and miscanthus do not need to be replanted each year and they do not require special care and high maintenance to grow. On the other hand, agricultural biomass residues (corn stover, wheat straw, rice husk, crop peels, pulps, etc.) as promising low-cost feedstocks since they do not need additional land for biomass growth and the land used for agriculture belongs to these types of biomass materials. Forest biomass are also large source of materials for biofuels and other value-added products production. However, high costs of their harvesting and transportation limit their use. In addition to the advantage and disadvantage listed above, different sources of biomass feedstocks do not have same composition, uniform size and shape, etc., that considerable affect efficiency of conversion processes for a specific product. Therefore, biomass feedstocks for a bio-refinery needs to be standardized.

3. Conclusions

Biomass materials have some challenges that need to be overcome for their fully utilization for biofuel and other useful products. Availability, abundance, and requirements for growth, growth rate, etc., parameters considerably affect the feedstock selection for value-added products. Besides, the content of cellulose, hemicellulose, and lignin in biomass materials and accessibility of these fractions in the biomass structures play significant roles in biomass dissolution and biofuel production from the hydrolysate. Biomass material should be used in densified forms to overcome moisture, storage and handling problems. Biomass feedstocks delivered to a bio-refinery from different sources should be standardized for an effective conversion of a specific product.

Author details

Sibel Irmak
Biological Systems Engineering, Industrial Agricultural Products Center,
University of Nebraska-Lincoln, Lincoln, NE, USA

*Address all correspondence to: sibel.irmak@unl.edu

References

[1] Johansson J, Liss J, Gullberg T, Bjorheden R. Transport and handling of forest energy bundles-advantages and problems. Biomass and Bioenergy. 2006;**30**:334-341. DOI: 10.1016/j.biombioe.2005.07.012

[2] Balan V. Current challenges in commercially producing biofuels from lignocellulosic biomass. ISRN Biotechnology. 2014;**2014**:1-31. Article ID: 463074. http://dx.doi.org/10.1155/2014/463074

[3] Sokhansanj S, Fenton J. Cost benefits of biomass supply and pre-processing. A BIOCAP Research Integration Program Synthesis Paper. Canada: BIOCAP Foundation; 2006

[4] Kaliyan N, Morey RV, White MD, Tiffany DG. A tub-grinding/roll-press compaction system to increase biomass bulk density: Preliminary study. In: 2009 ASABE Annual International Meeting, Presentation Paper Number: 096658, Nevada. June 21–June 24; 2009

[5] Sokhansanj S, Turhollow AF. Biomass densification—Cubing operations and costs for corn stover. Applied Engineering in Agriculture. 2004;**20**(4):495-499

[6] Kaliyan N, Morey RV. Factors Affecting Strength and Durability of Densified Products. ASABE Paper No. 066077. St. Joseph, MI: ASABE; 2006

[7] Bali G, Khunsupat R, Akinosho H, Payyavula RS, Samuel R, Tuskan GA, et al. Characterization of cellulose structure of Populus plants modified in candidate cellulose biosynthesis genes. Biomass and Bioenergy. 2016;**94**:146-154. DOI: 10.1016/j.biombioe.2016.08.013

[8] Meng X, Pu Y, Yoo CG, Li M, Bali G, Park DY, et al. An in-depth understanding of biomass recalcitrance using natural poplar variants as the feedstock. ChemSusChem. 2017;**10**:139-150. DOI: 10.1002/cssc.201601303

[9] Capanema EA, Balakshin MY, Kadla JF. Quantitative characterization of a hardwood milled wood lignin by NMR spectroscopy. Journal of Agricultural and Food Chemistry. 2005;**53**:9639-9649. DOI: 10.1021/jf0515330

[10] Lewis NG, Yamamoto E. Lignin: Occurrence, biogenesis and biodegradation. Annual Review of Plant Physiology and Plant Molecular Biology. 1990;**41**:455-496. DOI: 10.1146/annurev.pp.41.060190.002323

[11] Barakat A, Vries de H, Rouau X. Dry fractionation process as an important step in current and future lignocellulose biorefineries: A review. Bioresource Technology. 2013;**134**:362-373. DOI: 10.1016/j.biortech.2013.01.169

[12] Irmak S, Meryemoglu B. Comparison of perennial grasses and corn based agricultural biomass materials for high-yielding hydrogen gas production. Transaction of the ASABE. 2017;**60**(3):601-606. DOI: 10.13031/trans.12088

[13] Puls J. Chemistry and biochemistry of hemicelluloses. Relationship between hemicellulose structure and enzymes required for hydrolysis. Macromolecular Symposia. 1997;**120**:183-196. DOI: 10.1002/masy.19971200119

[14] Alvira P, Tomás-Pejó E, Ballesteros M, Negro M. Pretreatment technologies for an efficient bioethanol production process based on enzymatic hydrolysis: A review. Bioresource Technology. 2010;**101**:4851-4861. DOI: 10.1016/j.biortech.2009.11.093

[15] Galbe M, Zacchi G. Pretreatment of lignocellulosic materials for efficient bioethanol production. In: Scheper T,

Olssen L, editors. Biofuels. Vol. 2007. Berlin, Heidelberg, Germany: Springer; 2007. pp. 41-65. DOI: 10.1007/978-3-540-73651-6

[16] Mosier N, Wyman C, Dale B, Elander R, Lee Y, Holtzapple M, et al. Features of promising technologies for pretreatment of lignocellulosic biomass. Bioresource Technology. 2005;**96**:673-686

[17] Brennan AH, Hoagland W, Schell DJ. High temperature acid-hydrolysis of biomass using an engineering scale plug flow reactor: Results of low solids testing. Biotechnology and Bioengineering Symposium. 1986;**17**:53

[18] Wijaya YP, Putra RD, Widyaya VT, Ha JM, Suh DJ, Kim CS. Comparative study on two-step concentrated acid hydrolysis for the extraction of sugars from lignocellulosic biomass. Bioresource Technology. 2014;**164**:221-231. DOI: 10.1016/j.biortech.2014.04.084

[19] Sainio T, Turku I, Heinonen J. Adsorptive removal of fermentation inhibitors from concentrated acid hydrolyzates of lignocellulosic biomass. Bioresource Technology. 2011;**102**:6048-6057. DOI: 10.1016/j.biortech.2011.02.107

[20] Kanchanalai P, Temani G, Kawajiri Y, Realff MJ. Reaction kinetics of concentrated-acid hydrolysis for cellulose and hemicellulose and effect of crystallinity. BioResources. 2006;**11**:1672-1689. DOI: 10.15376/biores.11.1.1672-1689

[21] Wingren A, Galbe M, Zacchi G. Techno-economic evaluation of producing ethanol from softwood: Comparison of SSF and SHF and identification of bottlenecks. Biotechnology Progress. 2003;**19**: 1109-1117. DOI: 10.1021/bp0340180

[22] Galbe M, Zacchi G. A review of the production of ethanol from softwood. Applied Microbiology and Biotechnology. 2002;**59**:618-628. DOI: 10.1007/s00253-002-1058-9

[23] Sun Y, Cheng J. Hydrolysis of lignocellulosic materials for ethanol production: A review. Bioresource Technology. 2002;**83**:1-11. DOI: 10.1016/S0960-8524(01)00212-7

[24] Kruse A, Dinjus E. Hot compressed water as reaction medium and reactant: Properties and synthesis reactions. Journal of Supercritical Fluids. 2007;**39**:362-380. DOI: 10.1016/j.supflu.2006.03.016

[25] Herrero M, Cifuentes A, Ibanez E. Sub- and supercritical fluid extraction of functional ingredients from different natural sources: Plants, food by-products, algae and microalgae. Food Chemistry. 2006;**98**:136-148. DOI: 10.1016/j.foodchem.2005.05.058

[26] Ayala RS, Castro L. Continuous subcritical water extraction as a useful tool for isolation of edible essential oils. Food Chemistry. 2001;**75**:109-113. DOI: 10.1016/S0308-8146(01)00212-6

[27] Rogalinski T, Herrmann S, Brunner G. Production of amino acid from bovine serum albumin by continuous subcritical water hydrolysis. Journal of Supercritical Fluids. 2005;**36**:49-58. DOI: 10.1016/j.supflu.2005.03.001

[28] Khajavi SH, Ota S, Kimura Y, Adachi S. Kinetics of maltooligosaccharide hydrolysis in subcritical water. Journal of Agricultural and Food Chemistry. 2006;**54**:3663-3667. DOI: 10.1021/jf060117s

[29] Glasser L. Water, water, everywhere: Phase diagrams of ordinary water substance. Journal of Chemical Education. 2004;**81**:414-418. DOI: 10.1021/ed081p414

[30] Cooper JR, Dooley RB. Release on the ionization constant of H_2O. Switzerland: The International Association for the Properties of Water and Steam Lucerne; 2007. Available from: http://www.iapws.org/relguide/Ionization.pdf

[31] Irmak S, Ozturk I. Hydrogen rich gas production by thermocatalytic decomposition of kenaf biomass. International Journal of Hydrogen Energy. 2010;**35**:5312-5317. DOI: 10.1016/j.ijhydene.2010.03.081

[32] Ozturk I, Irmak S, Hesenov A, Erbatur O. Hydrolysis of kenaf (*Hibiscus cannabinus L.*) stems by catalytical thermal treatment in subcritical water. Biomass and Bioenergy. 2010;**34**:1578-1585. DOI: 10.1016/j.biombioe.2010.06.005

[33] Suchy M, Argyropoulos DS. Catalysis and activation of oxygen and peroxide delignification of chemical pulps: A review. American Chemical Society Symposium Series. 2001;**785**: 2-43. DOI: 10.1021/bk-2001-0785.ch001

[34] Northey RA. A review of lignin model compound reactions under oxygen bleaching conditions. American Chemical Society Symposium Series. 2001;**785**:44-60. DOI: 10.1021/bk-2001-0785.ch002

[35] Meryemoglu B, Hesenov A, Irmak S, Atanur OM, Erbatur O. Aqueous-phase reforming of biomass using various types of supported precious metal and raney-nickel catalysts for hydrogen production. International Journal of Hydrogen Energy. 2010;**35**:12580-12587. DOI: 10.1016/j.ijhydene.2010.08.046

Chapter 2

Different Pretreatment Methods of Lignocellulosic Biomass for Use in Biofuel Production

Muhammad Nauman Aftab, Irfana Iqbal, Fatima Riaz, Ahmet Karadag and Meisam Tabatabaei

Abstract

Lignocellulosic biomasses are carbon neutral and abundantly available renewable bioresource material available on earth. However, the main problem that hinders its frequent use is the tight bonding within its constituents that include cellulose, hemicellulose, and lignin. The selection of pretreatment process depends exclusively on the application. Various pretreatment processes are primarily developed and utilized in effective separation of these interlinked components to take maximum benefit from the constitutes of the lignocellulosic biomasses especially for the production of biofuel. The major pretreatment methods include physical, chemical, thermophysical, thermochemical, and biological approaches. Various aspects of these different pretreatment approaches are discussed in this chapter.

Keywords: pretreatment methods, lignocellulosic biomass, biofuel production

1. Introduction

We are living in a world of many challenges such as climate changes, polluted environment, resource depletion, and increasing demand for fuel. The use of oil reserves to fulfill our need of fuel has caused many drastic challenges from energy security to change in temperature. Rapid industrialization has increased the demand of petroleum products and consequently has raised the monopoly of few countries, which can manipulate petroleum price and create instability. This may also create environmental problems by emission of greenhouse gases and subsequently effect on climate change. The most important source of energy is petroleum that is largely used in transportation and industries; therefore, viability of liquid fuel is enhanced. As the environmental issues are growing, more research is being conducted to address the problems. The search for alternative source of petrol that is less costly with minimal environmental effects has become the center of attention. For instance, biomass is considered as a sustainable resource that can be utilized in large-scale production of biofuel that can be utilized as an alternative source of fuel and may present solution to environmental problems. Furthermore, relying on fossil fuel could be detrimental as it has been predicted of its depletion by 2050. The total annual primary production of biomass is over 100 billion tonnes of carbon per year, and the energy reserve per metric tonne of biomass is between 1.5E3 and 3E3 kW hours that is sufficient to cater the needs of the world energy requirements [1].

Bioenergy products like bioethanol, biohydrogen, and biodiesel can be obtained from lignocellulose biomass which is considerably large renewable bioresource and obtained from plants. The term "lignocellulosic biomass" is defined as lignin, cellulose, and hemicellulose that constitute the plant cell wall. Strong cross-linking associations are present between these components that cause hindrance in the breakdown of plant cell wall. Polysaccharides and lignin are cross-linked via ester and ether linkages [2–4]. Microfibrils that are formed by cellulose, hemicellulose, and lignin help in the stability of plant cell wall structure [5, 6].

Lignocellulose was first produced from food crop such as corn, oilseed, and sugarcane. But the use of edible feedstock for bioenergy products formation is being discouraged to prevent the rise in food competition. Thus, second-generation biofuels are obtained from plants wastes to avoid competition of land and water resources between energy crops and food crops. Currently, lignocellulose is being produced from wood residues, agricultural residues, food industry residue, grasses, domestic wastes, municipal solid wastes, and nonfood seeds [7–9]. The lignocellulose wastes (LCW) are largest renewable bioresource reservoir on earth that is being wasted as pre and postharvest agricultural wastes. Thus, many steps need to be adopted for use of these renewable resources for the production of bioenergy products. Recovery of many products like enzymes, methane, activated carbon, lipids, resins, methane, carbohydrates, surfactants, resins, organic acids, ethanol, amino acids, degradable plastic composites, biosorbents, biopesticides, and biopromoters can be achieved by utilizing LCW. The added benefits of using LCW besides recovery of different products are the removal of LCW waste from the environment. Also, utilization of LCW eliminates the use of food for bioethanol production. The US government has planned the production of 21 billion gallon of biofuels by 2022 [2, 5]. Biofuel production from lignocellulosic biomass reduces the emission of greenhouse gases.

Pretreatment brings physical, biological, and chemical changes to biomass structure; therefore, it is very important to consider the type of pretreatment. In order to break down the hindrance caused by strong association within the cell wall, pretreatment is an important step which can increase the availability of lignocellulosic biomass for cellulase enzymes, their digestibility, and product yield. Before subjection to enzymatic hydrolysis, pretreatment of biomass can increase the rate of hydrolysis by 3–10-fold. Pretreatment of LCW is not an easy step as it seems after the installation of power generator; pretreatment is the second most costly process at industrial level. In crystalline cellulose, the disruption of hydrogen bonds, cross-linked matrix disruption, and increase in porosity as well as surface area of cellulose are the three tasks that are performed via a suitable pretreatment methods. The outcome of pretreatment also differs due to the difference in the ratio of cell wall components [10, 11]. The option to use dilute acid pretreatment method is more effective against poplar tree bark or corn as compared to the same method used for sweet gum bark or cornstalks. Few requirements of an effective, efficient, and economically suitable pretreatment process that including use of cheap chemicals, very less consumption of chemicals, prevention of hemicellulose and cellulose from denaturation, minimal energy requirement and consumption, cost-effective size reduction process, and reactive cellulosic fiber production are the factors that need to be considered for pretreatment. There are several methods of pretreatment that can be divided into four categories, namely, chemical, physical, biological, and physiochemical pretreatment [12–15].

2. Physical methods

Pore size and surface area of lignocellulosic biomass can be increased, whereas crystallinity and degree of polymerization of cellulose can be decreased with the application of physical methods. Physical pretreatments include milling, sonication, mechanical extrusion, ozonolysis, and pyrolysis.

2.1 Milling

On the inherent ultrastructure of cellulose and degree of crystallinity, milling can be performed to render lignocelluloses more amenable to cellulases. Cellulases are enzyme that catalyze cellulose, but for the catalysis and best results, the substrate availability needs to be enhanced for optimized functioning of the enzymes. Before the subjection of the LCW to enzymatic hydrolysis, milling and size reduction of the lignocellulosic matter should be performed. Milling process has several types like ball milling, colloid milling, vibro-energy milling, hammer milling, and two-roll milling. For wet material, colloid mill, dissolver, and fibrillator are suitable, whereas for dry materials hammer mill, extruder, cryogenic mill, and roller mill are used. For both wet and dry material, ball milling can be used. For waster paper, hammer milling is the most suitable pretreatment option. Enzymatic degradation can be improved by milling as it reduces the degree of crystallinity and material size. Up to 0.2 mm reduction in particle size can be seen by milling and grinding. Reduction in particle size of biomass can be achieved up to a certain limit; beyond that limit reduction in particle size does not effect in the pretreatment procedure. Corn stover with small particle size, i.e., from 53 to 75 μm, is more productive as compared to large particle size corn stover ranging from 475 to 710 μm. The difference in particle size shows that productivity can significantly affect the pretreatment process. Ball milling causes a massive drop in crystallinity index from 4.9 to 74.2% which makes this process more suitable for saccharification of straw at mild hydrolytic conditions with more production of fermentable sugars [12, 16–18]. For better results of hydrolysis, milling can be used in combination with enzymatic hydrolysis. Mechanical action, mass transfer, and enzymatic hydrolysis can be achieved at the same time when two methods are combined. A number of ball beads in bill mill reactor play a crucial role in the α-cellulose hydrolysis, as less enzyme loading is required, and 100% rate of hydrolysis can be achieved in comparison to pretreatment of biomass that is carried without the use of milling procedure. Highest hydrolysis rate with high yield of reducing sugar was obtained when rice straw was put into fluidized bed opposed jet mill for fine grinding after cutting, steam explosion, and pulverization. For pretreatment of biomass, ball milling is an expensive option in terms of energy consumption, which is a huge disadvantage at industrial scale. Also, incapability of milling for removing lignin makes it a less suitable option as enzyme accessibility to the substrate is reduced in the presence of lignin. Reduction in crystallinity, degree of polymerization, and increase in surface area can be effected by the type of biomass, type of milling used for pretreatment, and duration of the milling process [19–21].

For improving digestibility and reducing crystallinity, vibratory ball milling is very effective. Low energy consumption has an important advantage of using wet disk milling which produces fibers that improve hydrolysis of cellulose, whereas hammer milling produces finer bundles. Due to this reason milling is not preferred when wet disk milling is available [22, 23]. Other study results of conventional ball and disk milling are compared. With the use of conventional ball milling, maximum yields of xylose and glucose were obtained, i.e., 54.3 and 89.4%, respectively [24]. Wet milling produces less yield, but it has the advantage of not producing inhibitors and very low energy consuming capability. An increase of 110% in enzymatic hydrolysis was achieved when wet milling was combined with alkaline pretreatment. Optimum parameters for wet milling pretreatment of corn stover were 10 mm diameter 20 steel balls, 1:10 solid-to-liquid ratio, 350 rpm/min speed, and 0.5 mm particle size [25] (**Figures 1** and **2**).

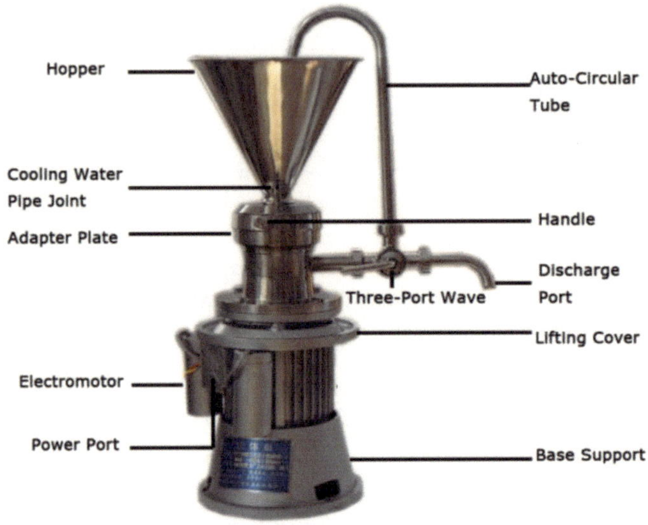

Figure 1.
Colloid milling (Pharmapproach.com).

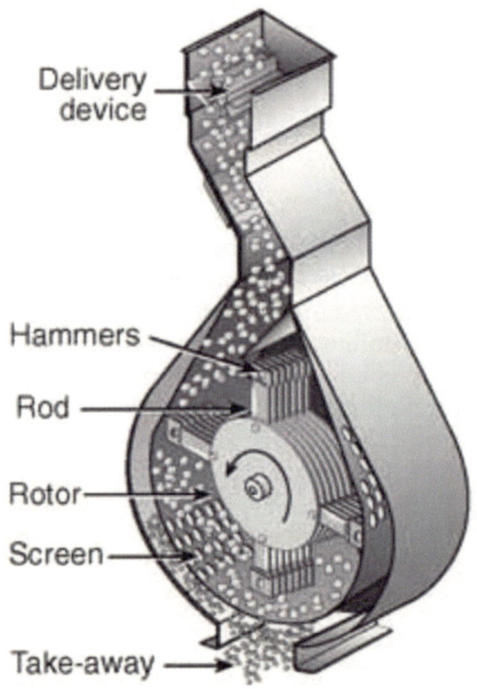

Figure 2.
Hammer milling (Solidswiki.com).

2.2 Microwave

Commonly used method for plant biomass pretreatment is microwave irradiation. This pretreatment method has several advantages that include ease of pretreatment, increased heating capacity, short processing time, minimal generation of inhibitors, and less energy requirement. Microwave irradiation in closed container was first reported in 1984 by team of researchers from Kyoto University,

Japan. They treated sugarcane bagasse, rice straw, and rice hulls with microwaves in the presence of water. The conditions used for microwave treatment include glass vessels of 50 mL, 2450 MHz energy, and 2.4 kW microwave irradiation [26]. Classical pretreatment methods were carried out at high pressure and temperatures. Chemical interactions between lignocellulosic material break as a result of high temperature, thus increasing substrate availability to the enzymes. Under high-pressure steam injection or indirect heat injection, high temperature between 160 and 250°C is provided to lignocellulosic material in conventional heating methods. However, in order to prevent temperature gradients, crushing of lignocellulosic material into small particles is needed. To avoid large temperature gradients, microwave is a good choice as it uniformly distributes heat which also avoids degradation of lignocellulosic material into humic acid and furfural. For effective degradation, microwave irradiation is combined with mild alkali treatment. Sugar yield of 70–90% from switch grass was obtained from alkali and irradiation combined pretreatment [27]. As microwave irradiation is performed at high temperature, therefore, closed containers are required to achieve high temperature. Three properties, namely, penetration, reflection, and absorbance are exhibited by microwave. Microwave passes through glass and plastic, absorbed by water and biomass, whereas microwaves are reflected by metals. Based on these properties, microwave reactors can be divided into two types, one that allows the passage of microwaves, whereas the other kind reflects the microwaves. Glass or plastic is the building material of the first type of microwave reactors, whereas the second types of reactors are composed of steel. Through quartz windows, microwaves can enter into the reactor as these are placed in the reactor. Closed, sealable, pressure-resistant glass tube container having gasket made up of Teflon can be used for the high temperature, i.e., 200°C, for microwave irradiation pretreatment. Sensors are used to control and ensure temperature inside the microwave. Teflon-coated sensors are a good choice because of the thermostability, corrosion-free nature, and zero absorbance properties. In a microwave oven, Teflon vessels are used by some scientists due to its advantageous properties [28, 29]. Normally vessel sizes vary from 100 mL to several hundred milliliters. A 650 mL vessel with 318 mm length, connected nitrogen bottle, gauges, and thermometers are installed on the top of the microwave that was designed by Chen and Cheng [30]. Besides the glass vessels and stainless steel tanks with temperature and pressure sensors, automatic controlling system for microwave input and mechanical stirrer are also used (**Figure 3**).

2.3 Mechanical extrusion

When materials that can pass through a defined cross section die, it appears out with the fixed definite profile. This is the extrusion process which is known for sugar recovery from biomass. Adaptability to modifications, no degradation products, controllable environment, and high throughput are few advantages related to mechanical extrusion pretreatment process. Single screw extruder and twin screw extruder are two types of extruders.

Single screw extruder is based on three screw elements, forward, kneading, and reverse. With the minimum shearing and mixing, bulk material of varying pitches and lengths can be transported by forward screw element. Prominent mixing and shearing effect is produced by kneading screw elements with weak forward conveying effect, whereas the use of immense mixing and shearing involves material that is pushed back by reverse screw elements. A screw configuration is defined by the arrangement of different stagger angels, lengths spacing, pitches, and positions. Twin screw extruder can accomplish multiple tasks at the same time like mixing, shearing, grinding, reaction, drying, and separation. High enzymatic

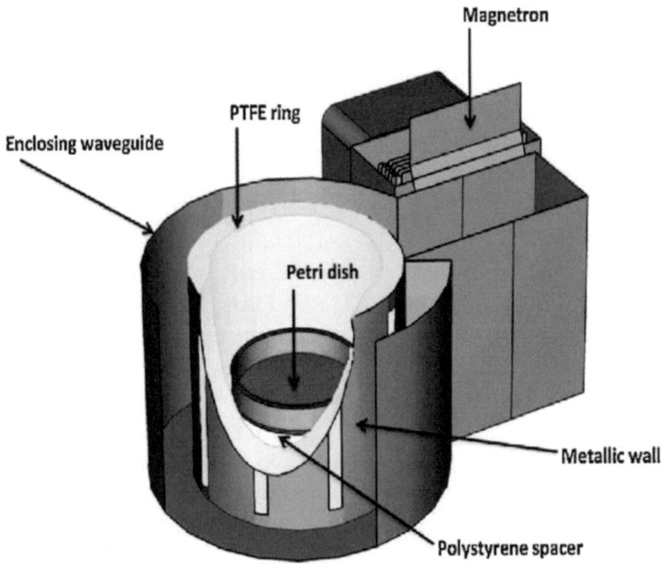

Figure 3.
Microwave irradiation (Researchgate.net).

hydrolysis rates are achieved by the use of single and twin screw extruders. Different parameters like speed of screw, temperature of barrel, and compression ratio can significantly affect recovery of sugars. Short-time extruders provide fast heat transfer, proper mixing, and increased shear. When material passed through the extruder barrel, structure of biomass is disturbed, exposing more surface for enzymatic hydrolysis [31–33]. During extrusion process, lignocellulosic material can be treated with alkali or acid in order to increase sugar recovery. Acidic treatment is less preferred than alkali because of the corrosion caused by acid to the extruder material. Corrosion problem can be solved by the use of AL6XN alloy for barrel fabrication and screws of extruder. With less carbohydrate degradation and role in the delignification, alkali treatment is suitable for lignocellulosic material. Sodium hydroxide is most commonly used to break ester linkages and solubilization of lignins and hemicelluloses. Alkali treatment can be applied by addition of alkali using volumetric pump into the extruder or by soaking the lignocellulosic material in alkali at room temperature [31, 34, 35] (**Figure 4**).

2.4 Pyrolysism

For the production of bio oil from biomass, process of pyrolysis is used. Pyrolysis is a thermal degradation of lignocellulosic biomass at very high temperature without the presence of oxidizing agent. At temperature ranging between 500 and 800°C, pyrolysis was performed. Rapid decomposition of cellulose resulted in the formation of products like pyrolysis oil and charcoal [36]. Based on temperature, pyrolysis pretreatment process is divided into fast and low pyrolysis. Certain factors affect the end products like biomass characteristics, reaction parameters, and type of pyrolysis. Due to high-value energy-rich product formation, easy transport management retrofitting, combustion, storage, and flexibility in utilization and marketing, thermal industries are adapting to the process of pyrolysis. Presence of oxygen and less temperature increase the efficiency of this process. A study on the bond cleavage rate of cellulose was carried out in the presence of nitrogen and oxygen. During the process of pyrolysis, breakage of 7.8×10^9 bonds/min/g cellulose

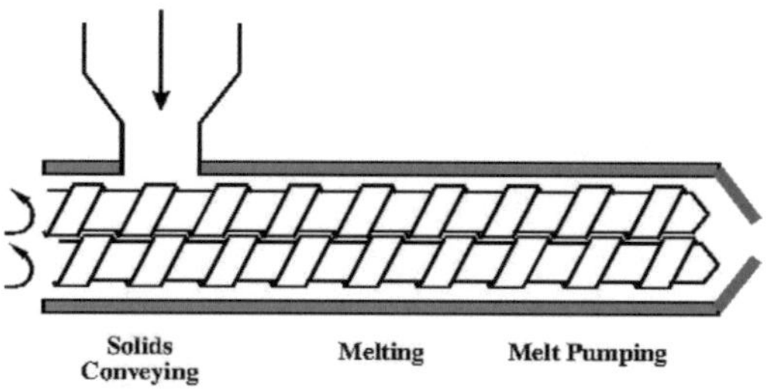

Figure 4.
Twin screw extruder (Researchgate.net).

in the presence of oxygen and breakdown of 1.7 × 10^8 bonds/min/g cellulose in the presence of nitrogen at 25°C were observed. In order to obtain more efficiency and results, microwave-assisted pyrolysis is preferred due to the microwave dielectric heating [37]. Thermochemical conversion of biomass into biofuels can be performed via three technologies, gasification, pyrolysis, and direct combustion [38]. Different yields of products from pyrolysis are due to different modes of pyrolysis. Bio oil is a mixture of polar organics and water. Pyrolysis is used where bio oil production is required. Fast pyrolysis in a controlled environment leads to the formation of liquid products (fuels). Torrefaction is an emerging technique which is also known as mild pyrolysis. It differs from pyrolysis with reference to thermochemical process that is carried out at temperature range between 200 and 300°C. Partial decomposition of biomass occurs in this process, and ultimate product obtained is terrified biomass. Whereas, in the process of pyrolysis, plant biomass is decomposed into vapor, aerosols, and char. Torrefaction has been categorized into two categories based on dry and wet torrefaction.

Dry torrefaction needs an inert environment and completely dry biomass and normal atmospheric pressure. Biochar is the major product in this type of biomass pretreatment. Hydrothermal carbonization and hydrothermal torrefaction are other terminologies used for wet torrefaction. Unlike dry torrefaction, pressurized vessel of water is used to carry out the pretreatment. Biomass used for wet torrefaction contains moisture content, but after torrefaction, a drying process is necessary in this type of torrefaction. A pressure between 1 and 250 MPa is required to carry out wet torrefaction. Biomass used during wet torrefaction pretreatment produces hydro-char as a main product [39].

2.5 Pulse electric field (PEF)

In this method, pores are created in the cell membrane due to which cellulose exposes to such agents that cause its breakdown by entering into the cell. High voltage ranging between 5.0 and 20.0 kV/cm is applied in a sudden burst to biomass for nano- to milliseconds. Sample was placed between two parallel plate electrodes, and the strength of electric field is given as $E = V/d$, where V and d are voltage and distance, respectively, between plate electrodes. Dramatic increase in mass permeability and tissue rapture occurred on the application of electric field. Electric pulses are applied, generally in the form of square waves or exponential decay. Setup of pulse electric field consisted of pulse generator, control system, data acquisition system, and material handling equipment [40, 41]. At ambient temperatures, the treatment can be performed at low energy. Another advantage of this treatment is the simple design of the instrument. Short duration of pulse time saves the effort and energy [42, 43]. Pulse electric field pretreatment was applied to pig manure and waste activated sludge by Author et al. [44]. As compared to untreated

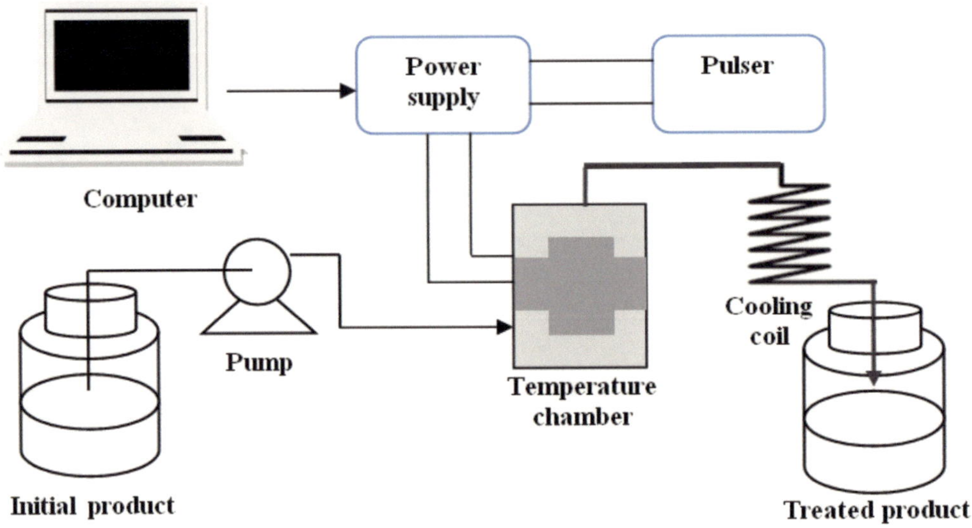

Figure 5.
Pulse electric field (Intechopen.com).

manure and sludge, 80% methane from manure and twofold increase in methane production from sludge were recorded in the study. A PEF system was designed and developed by Kumar et al. [45] that consisted of high-voltage power supply, switch circuit, a function generator, and sample holder. Neutral red dye was used to study the changes in the structure of cellulose by PEF pretreatment. Function generator drives the transistor present in the switching circuit; when pulse is applied by function generator to the switching circuit, switching circuit is turned on. Switching circuit is then transferred to the high voltage across the sample holder. So, by using function generator pulses of desired shape, width and high voltage can be applied to the sample. By using this setup, effects were observed on switch grass and wood. Results showed that at ≥8 kV/cm, switch grass showed high neutral red uptake. At low field strength, structural changes are less likely to occur. Electroporation through pulsed electric field is greatly affected by two parameters, pulse duration and electric field strength. Irreversible electroporation at >4 kV/cm with pulse duration in millisecond and ≥ 10 kV/cm with microsecond pulse duration was observed in *Chlorella vulgaris* which showed that pulse duration with a difference in micro- and milliseconds range can effect electroporation. Pulse electric field can increase hydrolysis rate by exposing cellulose to catalytic agents [40, 41, 46] (**Figure 5**).

3. Chemical pretreatments

3.1 Acid pretreatment

In this pretreatment, acids are used to pretreat lignocellulosic biomass. The generation of inhibitory products in the acid pretreatment renders it less attractive for pretreatment option. Furfurals, aldehydes, 5-hydroxymethylfurfural, and phenolic acids are the inhibitory compounds that are generated in huge amount in acid pretreatment. There are two types of acid treatments based on the type of end application. One treatment type is of short duration, i.e., 1–5 min, but high temperature > 180°C is used, and the second treatment type is of long duration, i.e., 30–90 min, and low temperature < 120°C is utilized. Due to hydrolysis by acid treatment, separate step of hydrolysis of biomass can be skipped, but to remove acid, washing is required before the fermentation of sugars [43, 47]. For acid pretreatment, such reactors are required that show resistance to corrosive, hazardous, and toxic acids; therefore, acid pretreatment is very expensive. Flow through, percolation, shrinking-bed, counter current rector, batch, plug flow are different types of rectors that have been developed. For enhancing economic feasibility of acid pretreatment, recovery of concentrated acid at the end of the treatment is an important step.

To treat lignocellulosic biomass, concentrated acids are also used. Most commonly used acids are sulfuric acid or hydrochloric acid. In order to improve the process of hydrolysis for releasing fermentable sugars from lignocellulosic biomass, acid pretreatment can be given. For poplar, switch grass, spruce, and corn stover, sulfuric acid pretreatment is commonly used. Reducing sugars of 19.71 and 22.93% were produced as a result of the acid pretreatment of Bermuda grass and rye, respectively. In percolation reactor, pretreatment of rice straw was carried out in two stages using aqueous ammonia and dilute sulfuric acid. When ammonia is used, 96.9% reducing sugar yield was obtained, while 90.8% yield was obtained in case of utilization of dilute acid. *Eulaliopsis binate* is a perennial grass and yielded 21.02% sugars, 3.22% lignin, and 3.34% acetic acid, and inhibitors in very less amount are produced when treated with dilute sulfuric acid [48, 49]. At 4 wt% concentration of sulfuric acid, pretreatment is preferred because of less cost and more effectiveness of the process. Dilute sulfuric acid causes biomass hydrolysis and then further breakdown of xylose into furfural is achieved. High temperature favors hydrolysis by dilute sulfuric acid [50]. Removal of hemicellulose is important to increase glucose yield from cellulose, and dilute sulfuric acid is very effective to achieve this purpose. It is necessary for an economical biomass conversion to achieve high xylan-to-xylose ratio. One-third of the total carbohydrate is xylan in most lignocellulosic materials. There are two types of dilute acid pretreatments, one is characterized by high temperature, continuous flow process for low solid loadings, and the other one is with low temperature, batch process and high solid loadings. Temperature and solid loadings for the first type are >160°C and 5–10%, respectively, and for the second type, temperature and solid loadings are around<160°C and 10–40%, respectively [51, 52].

Besides sulfuric acid and hydrochloric acid, other acids like oxalic acid and maleic acid are also used for the pretreatment of lignocellulosic biomass. Oxalic and maleic acids have high pKa value and solution pH as compared to sulfuric acid. Because of having two pKa values, dicarboxylic acids hydrolyze biomass more efficiently than sulfuric acid and hydrochloric acid. Other advantages include less toxicity to yeast, no odor, more range of pH and temperature for hydrolysis, and no hampering of glycolysis. Maleic acid has k_{hyd}/k_{deg}, due to which hydrolysis of cellulose to glucose is preferred over glucose breakdown. Effects of oxalic, sulfuric, and maleic acid pretreatment on biomass at the same combined severity factor (CSF) were determined [53]. The use of maleic acid produces high concentration of xylose and glucose as compared to oxalic acid.

3.2 Alkali pretreatment

Apart from acids, few bases are also used for pretreatment of biomass. Lignin contents greatly affected the result of alkaline treatment. As compared to other pretreatment methods, alkali treatment requires less pressure and temperature and ambient condition, but alkali pretreatment needs time in days and hours. Degradation of sugar in alkali treatment is less than that by acid treatment, and also the removal and recovery of caustic salt are possible and easy in case of alkali treatment. Ammonium, sodium, calcium, and potassium hydroxides are used for alkaline pretreatment, but among these sodium hydroxide is the most commonly used alkaline pretreatment agent, whereas calcium hydroxide is the cheapest yet effective among all other alkali agents for pretreatment. By neutralizing calcium with carbon dioxide, calcium can be recovered easily in form of insoluble calcium carbonate. Using lime kiln technology, calcium hydroxide can be regenerated. Apparatus required for alkali pretreatment is basically temperature controller, a tank, CO2 scrubber, water jacket, manifold for water and air, pump, tray, frame, temperature sensor, and heating element. The first step of pretreatment consists of making lime slurry with water. The next step is spraying of this slurried lime on biomass; after spray, store the biomass for hours or, in some case, days. Contact time can be reduced by increasing temperature [54–57]. Crystallinity index increases in lime pretreatment because of the removal of lignin and hemicellulose. Structural features resulting from lime pretreatment affect the hydrolysis of pretreated biomass. Correlation of three structural factors, viz., lignin, acetyl content and crystallinity, and enzymatic digestibility, was reported by Chang and Holtzapple [58]. He concluded that (1) regardless of crystallinity and acetyl content, in order to obtain high digestibility, extensive delignification is enough. (2) Parallel barriers to hydrolysis are removed by delignification and deacetylation. (3) Crystallinity does not affect ultimate sugar yield; however, it plays some role in initial hydrolysis. It is evident from these points that lignin content should be reduced to 10% and all acetyl groups should be removed by an effective pretreatment process. Thus in exposing cellulose to enzymes, alkaline pretreatment plays an important role. By increasing enzyme access to cellulose and hemicellulose and eliminating nonproductive adsorption sites, lignin removal can play its role in increasing effectiveness of enzyme.

3.3 Organosolv

Aqueous organic solvents like methanol, acetone, ethanol, and ethylene glycol are used in this method with specific conditions of temperature and pressure. Organosolv pretreatment is usually performed in the presence of salt catalyst, acid, and base. The biomass type and catalyst involved decide the temperature of pretreatment, and it can go up to 200°C. Lignin is a valuable product, and to extract lignin this process is used mainly. Cellulose fibers are exposed when lignin is removed, which leads to more hydrolysis. During organosolv pretreatment, fractions and syrup of cellulose and hemicellulose, respectively, are also produced. There are certain variable factors like catalyst type, temperature, and concentration of solvent and reaction time which affects the characteristics of pretreated biomass like crystallinity, fiber length, and degree of polymerization. Inhibitor formation is triggered by long reaction, high temperature, and acid concentrations [59, 60]. In a study by Park et al. [61], effect of different catalyst was checked for the production of ethanol and among sulfuric acid, sodium hydroxide, and magnesium sulfate, and sulfuric acid was found to be most effective in ethanol production, but for enhancing digestibility the use of sodium hydroxide is proven to be effective. Sulfuric acid is a good catalyst, but its toxicity and inhibitory nature make it less favorite. Organosolv is not a cost-effective pretreatment process because of the high cost of catalysts, but it can be made cost-effective by recovering and recycling of solvents. Solvent removal is important because its presence effects fermentation, microorganism growth, and enzymatic hydrolysis. There is added risk of handling such harsh organic solvents. Acid helps in hydrolysis and depolymerization of lignin. Upon cooling lignin is dissolved in phenol, and in the aqueous phase, sugars are present. Formasolv involving formic acid, H_2O, and hydrochloric acid is a type of organosolv in which lignin is soluble and at low temperature process can be carried out. For pretreatment with ethanosolv cellulose, hemicellulose and pure lignin can be recovered, but high pressure and temperatures are required when ethanosolv is used, and less toxic nature of ethanol as compared to other organosolv makes it favorite for pretreatment. Ethanosolv when used in pretreatment effects the enzymatic hydrolysis, so to prevent this low ethanol, water is used [62]. Recovery of ethanol and water reduces the overall cost of the pretreatment. For sugarcane bagasse Mesa et al. [63] used ethanosolv at 195°C for 60 min, and results showed formation of 29.1% sugars from 30% ethanol. Alcohol-based organosolv pretreatment is combined with ball milling by Hideno et al. [24] to pretreat Japanese cypress and observed a synergistic effect on digestibility. 50.1, 41.7, and 48.1% yield of organosolv pulping was obtained from ethylene glycol-water, acetic acid-water, and ethanol-water in a study done by Ichwan and Son [64]. Poplar wood chips were first treated with stream and then with organosolv to separate cellulose, lignin, and hemicellulose. About 88% hydrolysis of cellulose to glucose, 98% recovery of cellulose, and 66% increase in lignin extraction were reported by Panagiotopoulos et al. [65].

3.4 Ionic liquids

For the pretreatment of lignocellulose, scientist took a great interest in using ionic liquids, for decades. Ionic liquids containing cations or anions are a new class of solvents with high thermal stability and polarity, less melting point, and negligible vapor pressure [66, 67]. Normally large organic cations and small inorganic anions compose ionic liquids. Factors like degree of anion charge delocalization and cation structure significantly effect physical, biological, and chemical ionic liquid properties. Interactions between ionic liquids and biomass get affected by temperature, cations and anions, and time of pretreatment.

Ionic liquids actually compete for hydrogen bonding with lignocellulosic components, and in this competition disruption of network occurs. 1-Ethyl-3-methylimidazolium diethyl phosphate-acetate, 1-butyl-3-methylimidazolium-acetate, cholinium amino acids, cholinium acetate, 1-ethyl-3-methylimidazolium diethyl phosphate-acetate, 1-allyl-3-methylimidazolium chloride, and chloride are ionic liquids used for the treatment of rice husk, water hyacinth, rice straw, kenaf powder, poplar wood, wheat straw, and pine. Among other ionic liquids are imidazolium salts which are most commonly used [42]. 1-Butyl-3-methylimidazolium chloride is used for pretreatment by Dadi et al. [68] who observed a twofold increase in yield and rate of hydrolysis. For the pretreatment of rice straw, Liu and Chen [69] used 1-butyl-3-methylimidazolium chloride also known as (Bmim-Cl) and observed significant enhancement in the process of hydrolysis due to modifications in the structure of wheat straw by Bmim-Cl. Bmim-Cl played role in the reduction of polymerization and crystallinity. A twofold increase in hydrolysis yield from sugarcane bagasse was observed in a study by Kuo and Lee [70] as compared to untreated bagasse. 1-Ethyl-3-methylimidazolium-acetate is used in a study by Li et al. [71] for the pretreatment of switch grass in order to remove lignin at a temperature of 160°C for 3 hours. Results showed 62.9% lignin removal enhanced enzymatic digestibility, and reduced cellulose crystallinity was reported by Tan et al. [72] on palm tree pretreatment with 1-butyl-3-methylimidazolium chloride. Slight changes in composition of biomass occurred after ionic liquid pretreatments although significant changes were observed in the structure of biomass. Ionic liquid pretreatment is less preferred over other techniques because of high thermal and chemical stability, less dangerous conditions for processing, low vapor pressure of solvents, and retaining liquid state at wide range of temperature. Ionic liquids can be recycled easily and are non-derivatizing. Disadvantage of using ionic liquid pretreatment is that noncompatibility of cellulase and ionic liquids results in the unfolding and inactivation of cellulase. At less viscosity cellulose solubilizes at low temperature; that's why while using ionic liquids, viscosity is an important factor to be considered regarding the energy consumption of the whole process. High temperatures trigger more side reactions and negative side effects like reducing ionic liquid stability [73].

3.5 Ozonolysis

Ozone pretreatment is a great option for lignin content reduction in lignocellulosic biomass. In vitro digestibility of biomass is enhanced by the application of ozone pretreatment. Inhibitors are not formed in this pretreatment which is a great advantage because other chemical pretreatments produce toxic residues. In ozone pretreatment, ozone acts as an oxidant in order to break down lignin. Ozone gas is soluble in water and being a powerful oxidant, by breaking down lignin, releases less molecular weight, soluble compounds. Wheat straw, bagasse, cotton straw, green hay, poplar sawdust, peanut, and pine can be pretreated with ozone in order to degrade lignin and hemicellulose; however, only slight changes occur in hemicellulose, whereas almost no changes occur in cellulose. Ozonolysis apparatus consists of ozone catalytic destroyer, iodine trap used for testing efficiency of catalyst, oxygen cylinder, ozone generator, three-way valve, ozone UV spectrophotometer, pressure regulation valve, process gas humidifier, vent, and automatic gas flow control valve [40, 41, 74–76]. Moisture content hugely effects oxidization of lignin via ozone pretreatment as lignin oxidation decreases with increase in the moisture content of biomass. Ozone mass transfer is limited at less water concentration, which ultimately effects its reactivity with biomass. Longer residence time of ozone is caused by the blockage of pores by water film [77]. During ozonolysis, pH of water decreases because of the formation of organic acids. Alkaline media trigger delignification because it removes lignins that are bonded to carbohydrates [78, 79].

Biomass delignification is associated with the production of inhibitory compounds. Certain aromatic and polyaromatic compounds are produced as a result of delignification [80]. Structural changes in lignin are observed by Bule et al. [81] in a study; different lignin subunits showed aromatic opening and degradation of β-O-4 moieties in NMR analysis. How do aromatic structures of control- and ozone-pretreated samples differ? A spectrum showed a decrease in aromatic carbon signal concentration. Changes were observed in methoxy groups that suggest the breakdown of ester-linked structure. Different reactor designs are used for the ozone pretreatment of biomass, for example, batch reactor, Drechsel trap reactor, fixed bed reactor, rotatory bed reactor, and multilayer fixed bed reactor. Plug flow reactors are used by most researchers [82]. Heiske et al. [83] compared the characteristics of single layered and multiple layered bed reactors in order to improve the wheat straw conversion to methane. Straw with 16.2% lignin concentration was obtained from single layered reactor, whereas in multiple layered reactor, lignin concentration decreased up to 7.2% at the bottom layer. Due to wax degradation in ozone-pretreated wheat straw, production of fatty acid compounds is observed by Kádár et al. [84]. About 49% lignin degradation was observed when corn stover was pretreated with ozonolysis in a study by Williams [85].

4. Physicochemical pretreatment

4.1 Ammonia fiber expansion (AFEX)

AFEX technique belongs to the category of physicochemical pretreatment methods. In this low temperature process, concentrated ammonia (0.3–2 kg ammonia/

kg of dry weight) is used as a catalyst. Ammonia is added to biomass in a reactor of high pressure; after 5–45 min of cooking, pressure is released rapidly. Normally temperature around 90°C is used in this process. Ammonia can be recovered and reused because of its volatility. The principle of AFEX is similar to steam explosion. Apparatus for AFEX includes reactor, thermocouple well, pressure gauge, pressure relief valve, needle valve, sample cylinder, temperature monitor, and vent. Rate of fermentation is seen to be improved by AFEX pretreatment of various grasses and herbaceous crops. For treatment of alfalfa, wheat chaff and wheat straw AFEX technology is used. Hemicellulose and lignin cannot be removed by using AFEX technology; hence, small amount of material is solubilized only. Degradation of hemicellulose into oligomeric sugars and deacetylation occur during AFEX pretreatment which is the reason of hemicellulose insolubility. After AFEX pretreatment of Bermuda grass and bagasse, 90% hydrolysis of cellulose and hemicellulose was achieved. Effectiveness of AFEX pretreatment decreases with increase in the lignin content of biomass, for example, newspaper, woods, nutshells, and aspen chips. In case of AFEX pretreatment for newspaper and aspen chips, maximum hydrolysis yield was 40% and 50%, respectively. So for the treatment of biomass with high lignin content, AFEX pretreatment is not a suitable choice.

Ammonia recycle percolation (ARP) is another method that uses ammonia. Aqueous ammonia (10–15 wt %) is used in this method. With a fluid velocity of 1 cm/min and temperature of 150–170°C and residence time of 14 minutes, aqueous ammonia passes through biomass in this pretreatment, and ammonia is recovered afterwards. Under these conditions, ammonia reacts with lignin and causes the breakdown of lignin breakdown linkages. Liquid ammonia is used in AFEX technique whereas in ammonia recycle percolation method/technique, aqueous ammonia is used.

4.2 Steam explosion

In this method, high-pressure saturated steam is used to treat lignocellulosic biomass, and then suddenly pressure is reduced, due to which lignocellulosic biomass undergoes explosive decompression. Initiation temperature of steam explosion 160–260°C and 0.69–4.83 MPa pressure is provided for few seconds to minutes, and then lignocellulosic biomass is exposed and retained at atmospheric pressure for a period of time; this triggers hydrolysis of hemicellulose and at the end explosive decompression, terminated the whole process. Cellulose hydrolysis potential increases due to the cellulose degradation and lignin transformation caused by high temperature. During the steam explosion pretreatment, acid and other acids formed, which played their role in the hydrolysis of hemicellulose. Fragmentation of lignocellulosic material occurs due to turbulent material flow and rapid flashing of material to atmospheric pressure [86–88]. In steam explosion pretreatment, the use of sulfuric acid or carbon dioxide decreases time, temperature, and formation of inhibitory products and increases hydrolysis efficiency that ultimately leads to complete removal of hemicellulose. Steam explosion pretreatment is not that effective for pretreating soft woods; however, acid catalyst addition during the process is a prerequisite to make the substrate accessible to hydrolytic enzymes. By using steam, targeted temperature can be achieved to process the biomass without the need of excessive dilution. Sudden release of pressure quenches the whole process at the end and also lowers the temperature. Particulate structure of biomass gets opened by rapid thermal expansion which is used to terminate the reaction. Steam explosion gets affected by certain factors like moisture content, residence time, chip size, and temperature. By two ways optimal hydrolysis and solubilization of hemicellulose can be achieved; either use high temperature and short residence time or low temperature and high residence time. Low energy requirement is a great advantage of steam explosion pretreatment, whereas in mechanical pretreatment 70% more energy is required as compared to steam explosion pretreatment in order to obtain the same, reduced particle size. So far steam explosion pretreatment with addition of a catalyst is tested and came closest to scaling up at commercial level due to its cost-effectiveness. In Canada, at Iogen demonstration plant, steam explosion pretreatment is used at a pilot scale. For hardwood and agriculture residues, steam explosion pretreatment is a very effective pretreatment process.

4.3 Carbon dioxide explosion

Supercritical carbon dioxide explosion treatment falls in the category of physiochemical pretreatment. Scientists had tried to develop a process cheaper than ammonia fiber explosion and a process which would operate at temperature lower than stream explosion temperature. In this process, supercritical carbon dioxide is used that behaves like a solvent. Supercritical fluids are compressed at room temperature above its critical point. When carbon dioxide is dissolved in water, carbonic acid is formed which causes less corrosiveness due to its special features. During the process, carbon dioxide molecules enter into small pores of lignocellulosic biomass due to its small size. Carbon dioxide pretreatment is operated at low temperature which helped in prevention of sugar decomposition by acid. Cellulosic structure is disrupted when carbon dioxide pressure is released which ultimately increased the accessibility of the substrate to the cellulolytic enzymes for the process of hydrolysis [11, 40, 41, 43]. Dale and Moreira [89] used carbon dioxide pretreatment for alfalfa and observed 75% theoretical release of glucose. Zheng et al. [90] performed experiments to show comparison among ammonia explosion, steam pretreatment, and carbon dioxide pretreatment of recycled paper and sugarcane bagasse. The results showed that carbon dioxide explosion pretreatment is cost-effective than AFEX.

4.4 Liquid hot water (LHW)

Hot compressed water is another terminology used for this method of treatment. High temperature (160–220°C) and pressure (up to 5 MPa) are used in this type of pretreatment in order to maintain the liquid state of water. However, chemicals and catalysts are not used in liquid hot water pretreatment method [42]. In this method, water in liquid form remains in contact with lignocellulosic biomass for about 15 min. In this treatment pressure is used to prevent its evaporation, and sudden decompression or expansion in this pretreatment process is not needed. This method has proved to be very effective on sugarcane bagasse, wheat and rye straw, corncobs, and corn stover. Different terms like solvolysis, aqueous fractionation, aquasolv, and hydrothermolysis are used by different researchers to describe this pretreatment method [42, 60, 91]. Based on biomass flow direction and water flow direction into reactor, liquid hot water pretreatment can be performed in three different ways. The first method is co-current pretreatment, which is carried out by heating biomass slurry and water at high temperature, holding it for a controlled residence time at pretreatment conditions, and finally applying cool environment. The second method involves the countercurrent pretreatment that engages pumping of hot water against biomass at controlled conditions. The third method is the flow-through pretreatment, which can be carried out by the flow of hot water through lignocellulosic biomass which acts like a stationary bed.

To investigate the effect of liquid hot water pretreatment, a study was conducted by Abdullah et al. [92] that determined the different hydrolysis rates of cellulose and hemicellulose. Two steps of optimization of various conditions were considered. The first step was performed at less severity for hydrolyzing hemicellulose, whereas the second step was performed at high severity for cellulose depolymerization. Disadvantage of liquid hot water pretreatment is high energy consumption requirement for downstream process because of the involvement of large amount of water. However, the advantage of this process is that chemicals and catalysts are not required and no inhibitor is formed [60].

4.5 Wet oxidation

In this pretreatment method, oxygen/air and water or hydrogen peroxide is used to treat biomass at high temperatures (>120°C) for half an hour at 0.5–2 MPa pressure [11, 93]. This pretreatment method is also used for the treatment of waste water and soil remediation. This method has proven to be very effective for pretreatment of lignin enriched biomass. Certain factors like reaction time, oxygen pressure, and temperature effect the efficiency of wet oxidation pretreatment process. Water acts like acid at high temperature, so it induces hydrolysis reaction as hydrogen ion concentration increases with increase in temperature which ultimately decreases the pH value. Pentose monomers are formed as a result of hemicellulose breakdown in wet oxidation pretreatment, and oxidation of lignin occurs, but cellulose remains least affected. There are certain reports on the addition of alkaline peroxide or sodium carbonate. The addition of these chemical agents help in bringing down temperature reaction and reduce the formation of inhibitory compounds. Efforts to improve the degradation of hemicellulose at high temperature lead to the formation of inhibitory compounds like furfural and furfuraldehydes. However, amount of the production of inhibitors in wet oxidation pretreatment is certainly less than that of liquid hot water pretreatment or steam explosion method. There is extremely less possibility of using this process at industrial scale because of two reasons. One is the combustible nature of oxygen, and the other is the high cost of hydrogen peroxide used in the process [94].

4.6 SPORL treatment

SPORL stands for sulfite pretreatment to overcome recalcitrance of lignocellulose, and this technique is used for pretreatment of lignocellulosic biomass [95]. SPORL is performed in two steps. The first step involves treatment of biomass with magnesium or calcium sulfite for the removal of lignin and hemicellulose fractions. The second step involves the reduction in size of pretreated biomass via mechanical disk miler. Effect of SPORL pretreatment was studied by Zhu et al. [22, 23] on spruce chips by employing conditions like temperature 180°C, half an hour time duration, 8–10% bisulfite, and 1.8–3.7% sulfuric acid. By employing these conditions, more than 90% substrate was converted to cellulose when cellulase of 14.6 FPU and 22.5 CBU β-glucosidase was used in hydrolysis. Low-yield inhibitors like hydroxymethyl furfural (HMF) (0.5%) and furfural (0.1%) were produced during this process. These percentages are far less as compared to acid-catalyzed steam pretreatment of spruce. In another study, SPORL-pretreated Popular NE222, beetle-killed lodgepole pine, and Douglas fir were purified. Low contents of sulfur and molecular mass were obtained with high phenolic derivative production [96].

SPORL pretreatment on switch grass with temperature ranging between 163 and 197°C, 3–37 min time duration, 0.8–4.2% sulfuric acid dose, and 0.6–7.4% sodium sulfite dose was performed by Zhang et al. [97]. The results with enhanced digestibility by the removal of hemicellulose due to sulfonation and decreased hydrophobicity of lignin were obtained. SPORL yielded 77.3% substrate as compared to 68.1%

for dilute acid treatment and 66.6% through alkali pretreatment. When sodium sulfite, sodium hydroxide, and sodium sulfide were used in SPORL pretreatment of switch grass, an improved digestibility of switch grass was achieved. When SPORL treatment was applied with optimized conditions, 97% lignin and 93% hemicellulose were removed from water hyacinth, and 90% hemicellulose and 75% lignin were achieved for rice husk [98].

5. Biological pretreatment

Conventional methods for chemical and physical pretreatments require expensive reagents, equipment, and high energy. On the other hand, biological pretreatment requires live microorganisms for the treatment of lignocellulosic material, and this method is more environment friendly and consumes less energy. There are certain microorganism present in nature that exhibit cellulolytic and hemicellulolytic abilities. White-rot, soft-rot, and brown fungi are known for lignin and hemicellulose removal with a very little effect on cellulose. White rot is able to degrade lignin due to the presence of lignin degrading enzymes like peroxidases and laccases. Carbon and nitrogen sources are involved in the regulation of these degrading enzymes [41]. Cellulose is commonly attacked by brown rot, whereas white and soft rot target both lignin and cellulose contents of plant biomass. Commonly used white-rot fungi species are *Pleurotus ostreatus, Ceriporiopsis subvermispora, Ceriporia lacerata, Pycnoporus cinnabarinus, Cyathus cinnabarinus*, and *Phanerochaete chrysosporium. Basidiomycetes* species including *Bjerkandera adusta, Ganoderma resinaceum, Trametes versicolor, Fomes fomentarius, Irpex lacteus, Lepista nuda*, and *Phanerochaete chrysosporium* are also tested, and these species showed high efficiency for delignification [41, 99].

Pretreatment of wheat straw was studied by Hatakka [100]. The results showed 13% conversion of wheat straw into sugars by *Pleurotus ostreatus* in duration of 5 weeks, whereas *Phanerochaete sordida* and *Pycnoporus cinnabarinus* showed almost the same conversion rate but in less time. For degradation of lignin in woodchips and to prevent cellulose loss, cellulase-less mutant of fungus *Sporotrichum pulverulentum* was developed [101]. Delignification of Bermuda grass by white-rot fungi *Ceriporiopsis subvermispora* and *Cyathus stercoreus* was studied that resulted in 29–32 and 63–77% improvement in delignification [102]. During the secondary metabolism in fungus *P. chrysosporium*, two lignin degrading enzymes, lignin peroxidase and manganese-dependent peroxidase, are produced in response to carbon and nitrogen limitation. Extracellular filtrates of various white-rot fungi contain these two enzymes.

Author details

Muhammad Nauman Aftab[1*], Irfana Iqbal[2], Fatima Riaz[1], Ahmet Karadag[3] and Meisam Tabatabaei[4]

1 Institute of Industrial Biotechnology, Government College University, Lahore, Pakistan

2 Department of Zoology, Lahore College for Women University, Lahore, Pakistan

3 Faculty of Science, Bartın University, Bartın, Turkey

4 Agricultural Biotechnology Research Institute of Iran (ABRII), AREEO, Karaj, Iran

*Address all correspondence to: nauman535@yahoo.com

References

[1] Field CB, Behrenfeld MJ, Randerson JT, Falkowski P. Primary production of the biosphere: Integrating terrestrial and oceanic components. Science. 1998;**281**:237-240

[2] da Costa Sousa L, Chundawat SP, Balan V, Dale BE. "Cradle-to-grave" assessment of existing lignocellulose pretreatment technologies. Current Opinion in Biotechnology. 2009;**20**(3):339-347

[3] Himmel ME. Biomass recalcitrance: Engineering plants and enzymes for biofuels production. Science. 2007;**315**:804-807

[4] Lin Y, Tanaka S. Ethanol fermentation from biomass resources: Current state and prospects. Applied Microbiology and Biotechnology. 2006;**69**(6):627-642

[5] Mtui GYS. Recent advances in pretreatment of lignocellulosic wastes and production of value added products. African Journal of Biotechnology. 2009;**8**(8):1398-1415

[6] Xiao C, Bolton R, Pan WL. Lignin from rice straw kraft pulping: Effects on soil aggregation and chemical properties. Bioresource Technology. 2007;**98**(7):1482-1488

[7] Qi BC, Aldrich C, Lorenzen L, Wolfaardt GW. Acidogenic fermentation of lignocellulosic substrate with activated sludge. Chemical Engineering Communications. 2005;**192**(7-9):1221-1242

[8] Rodríguez G, Lama A, Rodríguez R, Jiménez A, Guillén R, Fernández-Bolaños J. Olive stone an attractive source of bioactive and valuable compounds. Bioresource Technology. 2008, September

[9] Roig A, Cayuela ML, Sánchez-Monedero MA. An overview on olive mill wastes and their valorisation methods. Waste Management. 2006;**26**(9):960-969

[10] Aden A, Ruth M, Ibsen K, Jechura J, Neeves K. Lignocellulosic Biomass to Ethanol Process Design and Economics Utilizing Co-Current Dilute Acid Prehydrolysis and Enzymatic Hydrolysis for Corn Stover. 2002

[11] Haghighi Mood S, Hossein Golfeshan A, Tabatabaei M, Salehi Jouzani G, Najafi GH, Gholami M, et al. Lignocellulosic biomass to bioethanol, a comprehensive review with a focus on pretreatment. Renewable and Sustainable Energy Reviews. 2013;**27**:77-93

[12] Taherzadeh MJ, Karimi K. Pretreatment of lignocellulosic wastes to improve ethanol and biogas production: A review. International Journal of Molecular Sciences. 2008;**9**

[13] Wyman C. Biomass, Pretreatment of Handbook on Bioethanol: Production and Utilization. Washington DC: Tylor & Francis; 1996

[14] Torget R, Himmel ME, Grohmann K. Dilute sulfuric acid pretreatment of hardwood bark. Bioresource Technology. 1991

[15] Fan LT, Lee Y-H, Gharpuray MM. The nature of lignocellulosics and their pretreatments for enzymatic hydrolysis. Microbial Reactions. 1982;**23**:157-187

[16] Bondy SC, Guo-Ross SX, Pien J. Mechanisms underlying the aluminum-induced potentiation of the pro-oxidant properties of transition metals. Neurotoxicology. 1998;**19**(1):65-71

[17] Sidiras DK, Koukios EG. Acid saccharification of ball-milled straw. Biomass. 1989;**19**(4):289-306

[18] Walpot JI. Enzymatic hydrolysis of waste paper. Conservation & Recycling. 1986;**9**(1):127-136

[19] Muller CD, Abu-Orf M, Novak JT. Application of mechanical shear in an internal-recycle for the enhancement of mesophilic anaerobic digestion. Water Environment Research. 2007;**79**(3):297-304

[20] Jin S, Chen H. Superfine grinding of steam-exploded rice straw and its enzymatic hydrolysis. Biochemical Engineering Journal. 2006;**30**(3):225-230

[21] Mais U, Esteghlalian AR, Saddler JN, Mansfield SD. Enhancing the enzymatic hydrolysis of cellulosic materials using simultaneous ball milling. Applied Biochemistry and Biotechnology: Part A Enzyme Engineering and Biotechnology. 2002;**98-100**:815-832

[22] Zhu JY, Pan XJ, Wang GS, Gleisner R. Sulfite pretreatment (SPORL) for robust enzymatic saccharification of spruce and red pine. Bioresource Technology. 2009;**100**(8):2411-2418

[23] Zhu JY, Wang GS, Pan XJ, Gleisner R. Specific surface to evaluate the efficiencies of milling and pretreatment of wood for enzymatic saccharification. Chemical Engineering Science. 2009;**64**(3):474-485

[24] Hideno A, Kawashima A, Endo T, Honda K, Morita M. Ethanol-based organosolv treatment with trace hydrochloric acid improves the enzymatic digestibility of japanese cypress (*Chamaecyparis obtusa*) by exposing nanofibers on the surface. Bioresource Technology. 2013;**132**:64-70

[25] Lin Z, Huang H, Zhang H, Zhang L, Yan L, Chen J. Ball milling pretreatment of corn stover for enhancing the efficiency of enzymatic hydrolysis. Applied Biochemistry and Biotechnology. 2010;**162**(7):1872-1880

[26] Azuma JI, Tanaka F, Koshijima T. Enhancement of enzymatic susceptibility of lignocellulosic wastes by microwave irradiation. Wood Research. 1985;**71**:13-24

[27] Hu Z, Wen Z. Enhancing enzymatic digestibility of switchgrass by microwave-assisted alkali pretreatment. Biochemical Engineering Journal. 2008;**38**(3):369-378

[28] Komolwanich T, Tatijarern P, Prasertwasu S, Khumsupan D, Chaisuwan T, Luengnaruemitchai A, et al. Comparative potentiality of Kans grass (*Saccharum spontaneum*) and Giant reed (*Arundo donax*) as lignocellulosic feedstocks for the release of monomeric sugars by microwave/chemical pretreatment. Cellulose. 2014;**21**(3):1327-1340

[29] Azuma J, Tanaka F, Koshijima T. Enhancement of enzymatic susceptibility of lignocellulosic wastes by microwave irradiation. Journal of Fermentation Technology. 1984;**62**:377-384

[30] Chen W, Cheng N. Disruption of Sugarcane Bagasse Lignocellulosic Structure by Means of Dilute. Elsevier; 2011

[31] Zheng J, Rehmann L. Extrusion pretreatment of lignocellulosic biomass: A review. International Journal of Molecular Sciences. 2014;**15**(10):18967-18984

[32] Karunanithy C, Muthukumarappan K. Optimization of switchgrass and extruder parameters for enzymatic hydrolysis using response surface methodology. Industrial Crops and Products. 2011;**33**(1):188-199

[33] Amalia Kartika I, Pontalier PY, Rigal L. Twin-screw extruder for oil processing of sunflower seeds: Thermo-mechanical pressing and solvent extraction in a single step. Industrial Crops and Products. 2010;**32**(3):297-304

[34] Morrison IM. Influence of chemical and biological pretreatments on the degradation of lignocellulosic material by biological systems. Journal of the Science of Food and Agriculture. 1988;**42**(4):295-304

[35] Morrison IM. Changes in the biodegradability of ryegrass and legume fibres by chemical and biological pretreatments. Journal of the Science of Food and Agriculture. 1991;**54**(4):521-533

[36] Kilzer FJ, Broido A. Speculations on the nature of cellulose pyrolysis. Pyrodynamics. 1965;**2**:151-163

[37] Kappe CO. Controlled microwave heating in modern organic synthesis. Angewandte Chemie. 2004;**43**(46):6250-6284

[38] Yin C, Kær SK, Rosendahl L, Hvid SL. Co-firing straw with coal in a swirl-stabilized dual-feed burner: Modelling and experimental validation. Bioresource Technology. 2010;**101**(11):4169-4178

[39] Chen Z, Wang M, Ren Y, Jiang E, Jiang Y, Li W. Biomass torrefaction: A promising pretreatment technology for biomass utilization. IOP Conference Series: Earth and Environmental Science. 2018;**113**(1)

[40] Kumar P, Barrett DM, Delwiche MJ, Stroeve P. Methods for pretreatment of lignocellulosic biomass for efficient hydrolysis and biofuel production. Industrial and Engineering Chemistry Research. 2009;**48**(8):3713-3729

[41] Kumar P, Barrett DM, Delwiche MJ, Stroeve P. Methods for pretreatment of lignocellulosic biomass for efficient hydrolysis and biofuel production. Industrial and Engineering Chemistry Research. 2009

[42] Kumar AK, Sharma S. Recent updates on different methods of pretreatment of lignocellulosic feedstocks: A review. Bioresources and Bioprocessing. 2017;**4**(1)

[43] Kumar AK, Sharma S. Recent updates on different methods of pretreatment of lignocellulosic feedstocks: A review. Bioresources and Bioprocessing. 2017

[44] Author M, Salerno MB, Lee H-S, Parameswaran P, Rittmann BE. Using a pulsed electric field as a pretreatment for improved biosolids digestion and using a pulsed electric field as a pretreatment for improved biosolids digestion and methanogenesis. Rittmann Source Water Environ. Res. Biol. Treat. 2009;**81**(8):831-839

[45] Kumar P, Barrett DM, Delwiche MJ, Stroeve P. Pulsed electric field pretreatment of switchgrass and wood chip species for biofuel production. Industrial and Engineering Chemistry Research. 2011;**50**(19):10996-11001

[46] Luengo E, Martínez JM, Coustets M, Álvarez I, Teissié J, Rols M-P, et al. A comparative study on the effects of millisecond- and microsecond-pulsed electric field treatments on the permeabilization and extraction of pigments from *Chlorella vulgaris*. The Journal of Membrane Biology. 2015;**248**(5):883-891

[47] Sassner P, Mårtensson CG, Galbe M, Zacchi G. Steam pretreatment of H_2SO_4-impregnated salix for the production of bioethanol. Bioresource Technology. 2008;**99**(1):137-145

[48] Tang J, Chen K, Huang F, Xu J, Li J. Characterization of the pretreatment liquor of biomass from the perennial grass, *Eulaliopsis binata*, for the production of dissolving pulp. Bioresource Technology. 2013;**129**:548-552

[49] Kim Y, Yu A, Han M, Choi GW, Chung B. Enhanced enzymatic

saccharification of barley straw pretreated by ethanosolv technology. Applied Biochemistry and Biotechnology. 2011;**163**(1):143-152

[50] Mosier N, Wyman C, Dale B, Elander R, et al. Features of promising technologies for pretreatment of lignocellulosic biomass. Bioresource Technology. 2005;**96**(6):673-686

[51] Esteghlalian A, Hashimoto AG, Fenske JJ, Penner MH. Modeling and optimization of the dilute-sulfuric-acid pretreatment of corn stover, poplar and switchgrass. Bioresource Technology. 1997;**59**(2-3):129-136

[52] Brennan W, Hoagland W, Schell DJ, Scott CD. High temperature acid hydrolysis of biomass using an engineering-scale plug flow reactor: Results of low solids testing. Biotechnology and Bioengineering Symposium. 1986;**17**:53-70

[53] Lee JW, Jefferies TW. Efficiencies of acid catalysts in the hydrolysis of lignocellulosic biomass over a range of combined severity factors. Bioresource Technology. 2011

[54] Elshafei AM, Vega JL, Klasson KT, Clausen EC, Gaddy JL. The saccharification of corn stover by cellulase from *Penicillium funiculosum*. Bioresource Technology. 1991

[55] Kim S, Holtzapple MT. Effect of structural features on enzyme digestibility of corn stover. Bioresource Technology. 2006;**97**(4):583-591

[56] Lee D, Yu AHC, Wong KKY, Saddler JN. Evaluation of the enzymatic susceptibility of cellulosic substrates using specific hydrolysis rates and enzyme adsorption. Applied Biochemistry and Biotechnology. 1994;**45-46**(1):407-415

[57] MacDonald DG, Bakhshi NN, Mathews JF, Roychowdhury A, Bajpai P, Moo-Young M. Alkali treatment of corn stover to improve sugar production by enzymatic hydrolysis. Biotechnology and Bioengineering. 1983;**25**(8):2067-2076

[58] Chang VS, Holtzapple MT. Twenty-First Symposium on Biotechnology for Fuels and Chemicals. Fundamental Factors Affecting Biomass Enzymatic Reactivity. Totowa, NJ, United States: Humana Press; 2000

[59] Alriols M, Tejado A, Blanco M, et al. Palm oil tree residues as raw material for cellulose, lignin and hemicelluloses production by ethylene glycol pulping process. Chemical Engineering Journal. 2009;**148**(1):106-114

[60] Sparling R, Berlin A, Agbor VB, Cicek N, Sparling R, Berlin A, et al. Biomass pretreatment: Fundamentals toward application. Biotechnology Advances. 2011

[61] Park N, Kim HY, Koo BW, Yeo H, Choi IG. Organosolv pretreatment with various catalysts for enhancing enzymatic hydrolysis of pitch pine (*Pinus rigida*). Bioresource Technology. 2010;**101**(18):7046-7053

[62] Huijgen WJJ, Van der Laan RR, Reith JH. Modified organosolv as a fractionation process of lignocellulosic biomass for co-production of fuels and chemicals. Biomass. 2008:2-6

[63] Mesa L, González E, Cara C, et al. The effect of organosolv pretreatment variables on enzymatic hydrolysis of sugarcane bagasse. Chemical Engineering Journal. 2011

[64] Ichwan M, Son TW. Study on organosolv pulping methods of oil palm biomass. In: 2nd International Seminar on Chemistry 2011. 2011. pp. 364-370

[65] Panagiotopoulos IA, Chandra RP, Saddler JN. A two-stage pretreatment approach to maximise sugar yield

and enhance reactive lignin recovery from poplar wood chips. Bioresource Technology. 2013;**130**:570-577

[66] Behera S, Arora R, Nandhagopal N, Kumar S. Importance of chemical pretreatment for bioconversion of lignocellulosic biomass. Renewable and Sustainable Energy Reviews. 2014;**36**:91-106

[67] Zavrel M, Bross D, Funke M, Büchs J, Spiess AC. High-throughput screening for ionic liquids dissolving (ligno-) cellulose. Bioresource Technology. 2009;**100**(9):2580-2587

[68] Dadi AP, Varanasi S, Schall CA. Enhancement of cellulose saccharification kinetics using an ionic liquid pretreatment step. Biotechnology and Bioengineering. 2006;**95**(5):904-910

[69] Liu L, Chen H. Enzymatic hydrolysis of cellulose materials treated with ionic liquid [BMIM]Cl. Chinese Science Bulletin. 2006;**51**(20):2432-2436

[70] Kuo CH, Lee CK. Enhanced enzymatic hydrolysis of sugarcane bagasse by N-methylmorpholine-N-oxide pretreatment. Bioresource Technology. 2009;**100**(2):866-871

[71] Li C, Knierim B, Manisseri C, Arora R, et al. Comparison of dilute acid and ionic liquid pretreatment of switchgrass: Biomass recalcitrance, delignification and enzymatic saccharification. Bioresource Technology. 2010

[72] Tan HT, Lee KT, Mohamed AR. Pretreatment of lignocellulosic palm biomass using a solvent-ionic liquid [BMIM]Cl for glucose recovery: An optimisation study using response surface methodology. Carbohydrate Polymers. 2011;**83**(4):1862-1868

[73] Mäki-Arvela P, Anugwom I, Virtanen P, Sjöholm R, Mikkola JP. Dissolution of lignocellulosic materials and its constituents using ionic liquids—A review. Industrial Crops and Products. 2010

[74] Quesada J, Rubio M, Gómez D. Ozonation of lignin rich solid fractions from corn stalks. Journal of Wood Chemistry and Technology. 1999;**19**(1):115-137

[75] Vidal PF, Molinier J. Ozonolysis of lignin—Improvement of in vitro digestibility of poplar sawdust. Biomass. 1988;**16**(1):1-17

[76] Ben-Ghedalia D, Miron J. The effect of combined chemical and enzyme treatments on the saccharification and in vitro digestion rate of wheat straw. Biotechnology and Bioengineering. 1981;**23**(4):823-831

[77] Mamleeva NA, Autlov SA, Bazarnova NG, Lunin VV. Delignification of softwood by ozonation. Pure and Applied Chemistry. 2009;**81**(11):2081-2091

[78] Travaini R, Martín-Juárez J, Lorenzo-Hernando A, Bolado-Rodríguez S. Ozonolysis: An advantageous pretreatment for lignocellulosic biomass revisited. Bioresource Technology. 2015;**199**:2-12

[79] Chen CL, Chang HM, Gratzl JS, Pan GY. Studies an ozone bleaching. I. the effect of pH, temperature, buffer systems and heavy metal-ions on stability of ozone in aqueous solution. Journal of Wood Chemistry and Technology. 1984;**4**(3):367-387

[80] Travaini R, Otero MDM, Coca M, Da-Silva R, Bolado S. Sugarcane bagasse ozonolysis pretreatment: Effect on enzymatic digestibility and inhibitory compound formation. Bioresource Technology. 2013;**133**:332-339

[81] Bule MV, Gao AH, Hiscox B, Chen S. Structural modification of lignin and characterization of pretreated

wheat straw by ozonation. Journal of Agricultural and Food Chemistry. 2013;**61**(16):3916-3925

[82] Li C, Wang L, Chen Z, Li Y, Wang R, Luo X, et al. Ozonolysis pretreatment of maize stover: The interactive effect of sample particle size and moisture on ozonolysis process. Bioresource Technology. 2015;**183**:240-247

[83] Heiske S, Schultz-jensen N, Leipold F, Schmidt JE. Improving anaerobic digestion of wheat straw by plasma-assisted pretreatment. Journal of Atomic and Molecular Physics. 2013;**2013**:1-7

[84] Kádár Z, Schultz-Jensen N, Jensen JS, Hansen MAT, Leipold F, Bjerre AB. Enhanced ethanol production by removal of cutin and epicuticular waxes of wheat straw by plasma assisted pretreatment. Biomass and Bioenergy. 2015;**81**:26-30

[85] Williams KC. Subcritical water and chemical pretreatments of cotton stalk for the production of ethanol. 2006

[86] Jørgensen H, Kristensen JB, Felby C. Enzymatic conversion of lignocellulose into fermentable sugars: Challenges and opportunities. Biofuels, Bioproducts and Biorefining. 2007

[87] Li J, Henriksson G, Gellerstedt G. Lignin depolymerization/repolymerization and its critical role for delignification of aspen wood by steam explosion. Bioresource Technology. 2007;**98**(16):3061-3068

[88] SJB Duff WM. Bioconversion of forest products industry waste cellulosics to fuel ethanol: A review. Bioresource Technology. 1996

[89] Dale BE, Moreira MJ. Freeze-explosion technique for increasing cellulose hydrolysis. Biotechnology and Bioengineering Symposium. 1982;**12**:31-43

[90] Zheng Y, Lin HM, Tsao GT. Pretreatment for cellulose hydrolysis by carbon dioxide explosion. Biotechnology Progress. 1998;**14**(6):890-896

[91] Zhuang X, Wang W, Yu Q, Qi W, Wang Q, Tan X, et al. Liquid hot water pretreatment of lignocellulosic biomass for bioethanol production accompanying with high valuable products. Bioresource Technology. 2016;**199**:68-75

[92] Abdullah R, Ueda K, Saka S. Hydrothermal decomposition of various crystalline celluloses as treated by semi-flow hot-compressed water. Journal of Wood Science. 2014;**60**(4):278-286

[93] Varga E, Schmidt AS, Réczey K, Thomsen AB. Pretreatment of corn stover using wet oxidation to enhance enzymatic digestibility. Applied Biochemistry and Biotechnology: Part A Enzyme Engineering and Biotechnology. 2003;**104**(1):37-50

[94] Bajpai P. Pretreatment of Lignocellulosic Biomass. Singapore: Springer; 2016. pp. 17-70

[95] Xu H, Li B, Mu X. Review of alkali-based pretreatment to enhance enzymatic saccharification for lignocellulosic biomass conversion. Industrial and Engineering Chemistry Research. 2016

[96] Zhou H, Gleisner R, Zhu JY, Tian Y, Qiao Y. SPORL pretreatment spent liquors enhance the enzymatic hydrolysis of cellulose and ethanol production from glucose. Energy & Fuels. 2018;**32**(7):7636-7642

[97] Zhang DS, Yang Q, Zhu JY, Pan XJ. Sulfite (SPORL) pretreatment of switchgrass for enzymatic saccharification. Bioresource Technology. 2013;**129**:127-134

[98] Idrees M, Adnan A, Qureshi FA. Optimization of sulfide/sulfite pretreatment of lignocellulosic biomass for lactic acid production. BioMed Research International. 2013;**2013**:1-11

[99] Shi J. Microbial pretreatment of cotton stalks by *Phanerochaete chrysosporium* for bioethanol production [diss.]. N.C. State. 2007

[100] Hatakka AI. Pretreatment of wheat straw by white-rot fungi for enzymic saccharification of cellulose. European Journal of Applied Microbiology and Biotechnology. 1983;**18**(6):350-357

[101] Ander P, Eriksson KE. Selective degradation of wood components by white rot fungi. Physiologia Plantarum. 1977;**41**(4):239-248

[102] Akin D, Rigsby LL, Sethuraman A, Morrison WH, Gamble GR, Eriksson KE. Alterations in structure, chemistry, and biodegradability of grass lignocellulose treated with the white rot fungi *Ceriporiopsis subvermispora* and *Cyathus stercoreus*. Applied and Environmental Microbiology. 1995. Downloaded from http://aem.asm.org/ on September 12, 2013 by Nagaoka University of T. Am Soc Microbiol.

Chapter 3

Fungal Pretreatment of Lignocellulosic Materials

Najiah Nadir, Nur Liyana Ismail and Azlan Shah Hussain

Abstract

The biomass must be converted to fermentable carbohydrates through pretreatment process to break down the complex structure to its constituents prior to fermentation. For lignocellulosic materials, lignin moiety is extremely resistant to degradation because of hydrogen bond cross-linking between the cellulose and hemicellulose. Biological pretreatment using white-rot fungi are novel method and environmental-friendly as a method of biomass deconstruction as compared to other conventional means. These fungi can excrete ligninolytic enzymes to degrade lignin although the rate of deconstruction is slow. Hence, this chapter will focus on the fungal pretreatment or delignification process using white-rot fungi as it is an important step to increase the feedstock conversion.

Keywords: biomass conversion, lignocellulose, pretreatment, enzyme, white-rot fungi

1. Introduction

The depletion of fossil fuel energy sources causes much attention on biomass as the source of renewable energy or biofuel. There are three classifications of biofuel according to the feedstock source, which are first-, second- and third-generation biofuels. First-generation biofuels employ an edible biomass as feedstock while second-generation biofuels utilize numerous non-edible feedstock, ranging from lignocellulosic biomass to municipal solid wastes. Third-generation biofuels also exploit non-edible source but different feedstock such as algal biomass and gases [1].

The lignocellulosic biomass mainly composed of cellulose, hemicellulose and lignin. The conversion of lignocellulosic biomass requires a pretreatment process, which is one of the most important and expensive stages in bioenergy production. This process is performed to degrade and remove lignin from the biomass constituents and thus allows further manipulation of the valorizable portion of biomass, that is, increasing the yield of fermentable carbohydrates [2, 3]. Generally, the pretreatment process can be divided into four, that are physical, chemical, physicochemical, and biological methods [2, 4] as in **Table 1**. Subject to the pretreatment strategies, this process can reduce cellulose crystallinity, improve surface accessibility and decrease lignin content [3].

Biological pretreatment methods are performed by biological agents, either the microorganisms or enzymes excreted by the microorganisms. This method usually utilizes mild pressure and/or temperature and does not involve acid, alkali, or any

	Physical	Chemical	Physicochemical	Biological
Objective	Reduce particle size, increase surface area and reduce cellulose crystallinity	Hydrolyze lignin, hemicellulose and cellulose	Breakdown lignin-holocellulose linkages	Degrade lignin from holocellulose components
Type	• Milling • Grinding • Chipping • Freezing • Radiation	• Acid • Alkaline • Ionic liquid • Organosolv	• Steam explosion • Ammonia fiber expansion • CO_2 explosion • Liquid hot water • Wet oxidation	• Microbial consortium • Fungal • Enzymatic
Advantage	• Low environmental impact • Low dangerous chemical requirement • High effectiveness • Short process time • High uniformity and selectivity	• Less dangerous process condition • Lack of by-products degradation	• Less corrosiveness • Higher energy efficiency • Short process time	• Environmental friendly • No chemical requirement • Low energy consumption
Drawback	• High energy requirement • High cost	• Toxicity • Corrosiveness of equipment • Chemical recovery • Production of inhibitors • Long process time	• Chemical recovery and recycling • High operation cost • Formation of inhibitors	• Long process time • Large space requirement • Need continuous monitoring of microorganism growth

Table 1.
Pretreatment strategies of lignocellulosic biomass.

reactive species [2–4]. Since this pretreatment is conducted under mild conditions, it requires much lower energy input and the byproduct(s) would not hamper or inhibit hydrolysis process. Apart from that, there is no need for chemical recovery because no chemicals were employed [3, 5]. Due to these reasons, the biological pretreatment is an environmentally safe process [3–6]. However, the main issue is it consumes a long pretreatment time [4, 7].

This chapter discusses an overview of recent studies on fungal pretreatment using white-rot fungi and important parameters affecting the pretreatment process of lignocellulosic feedstock such as fungal strain, inoculum concentration and moisture content.

2. Fungal pretreatment

The biological pretreatment can be categorized into bacterial consortium, fungal treatments and enzymatic treatments [4, 8]. The commonly utilized microorganisms in this pretreatment of lignocellulosic biomass are filamentous fungi, which can be easily found in the environment such as ground, living plants and lignocellulose wastes [9]. Wood-decay fungi are classified into three main groups, which are white-, brown- and soft-rot fungi [10]. Among them, the most effective are basidiomycetes white-rot fungi because they have the capability to degrade lignin from the holocellulose (cellulose and hemicellulose) surface [2, 7, 9, 11] and cause white-rot on wood or trees, whereas brown- and soft-rot fungi degrade only minimal lignin [6]. Lignin is a polyaromatic polymer that gives rigidity to lignocellulose [7, 11]. Previous studies on three types of rot fungi were presented in **Table 2**.

2.1 White-rot fungi

White-rot fungi differ significantly in the relative rates at which they attack lignin and carbohydrates in woody or lignocellulosic tissues [6, 17]. They can be differentiated by their delignification mode, named as selective and non-selective delignification as can be seen in **Figure 1**. In selective delignification, mostly lignin and hemicellulose are degraded, while consuming a small amount of cellulose. However, for non-selective delignifiers, all three lignin, hemicellulose and cellulose are degraded almost equally [6, 18]. Even the number of non-selective white-rot fungi is greater than selective white-rot fungi [11], more than 1500 fungi species are selective delignifiers [6]. These fungi are favored for fungal pretreatment in recent researches to ensure a lignin-free and cellulose-rich biomass for next hydrolysis step [3, 7, 17] and enhance the biomass digestibility [3, 18]. Some of the white-rot fungi species were shown in **Figure 2**.

2.2 Enzymatic systems of white-rot fungi

The white-rot fungi play a major role in degrading woods in forest ecosystems [9]. These fungi have the ability to degrade lignocellulosic biomass during their growth in nature owing to the production of two enzymatic systems, which are hydrolytic system and oxidative ligninolytic system [7, 17]. In hydrolytic system, cellulases and hemicellulases are utilized to degrade holocellulose [17]. Non-selective white-rot fungi cause substantial cellulose loss because of their high cellulolytic and hemicellulolytic activity. Conversely, selective white-rot fungi excrete hemicellulolytic enzymes and employ hemicellulose-derived sugars as the main carbon sources [7].

Substrate	Fungi species	Effect	References
Wheat straw	*Ganoderma lobatum* (white-rot)	• Lignin, hemicellulose and cellulose degradation of 50.3, 18.1 and 21.4% • Sugar recovery increased by approximately 27.6%	[12]
	Gloeophyllum trabeum (brown-rot)	• 37.6 and 13.3% of hemicellulose and cellulose removal • Sugar recovery decreased by 10.9%	
Moso bamboo	*Phanerochaete chrysosporium* (white-rot)	• Higher degradability • On lignin over hemicellulose and cellulose	[13]
	G. trabeum (brown-rot)	• Preferential degradability on hemicellulose than lignin and cellulose	
Radiata pine	*Trametes versicolor* (white-rot)	• Loss of lignin at 16%, while both hemicellulose and cellulose at 5% each	[14]
	Stereum hirsutum (white-rot)	• Lignin degradation of 16%, whereas hemicellulose and cellulose of 9% individually	
	G. trabeum (brown-rot)	• Hemicellulose and glucans content reduced approximately to 5 and 3% • Mass loss ranged between 6 and 8% during the first month of biodegradation	[15]
Scots pine	*Daldinia concentrica* (soft-rot)	• 2.5% of weight loss after decayed for 2 months	[16]
	Xylaria acuta (soft-rot)	• Weight reduction of 12.4% after 2 months of incubation	

Table 2.
Previous studies on fungal pretreatment using three different types of rot fungi.

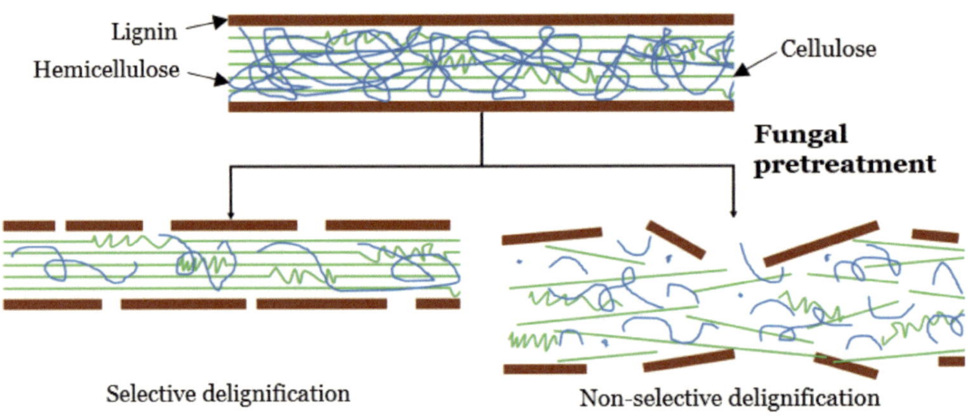

Figure 1.
Mechanism of fungal pretreatment using white-rot fungi on lignocellulosic materials.

Figure 2.
White-rot fungi of species (a) Abortiporus biennis, *(b)* Ceriporiopsis subvermispora, *(c)* Coriolopsis trogii, *(d)* Ganoderma applanatum, *(e)* Irpex lacteus *and (f)* Trametes versicolor *[19].*

The main enzymes in ligninolytic system to degrade lignin and open the phenyl rings are lignin peroxidase (LiP), manganese peroxidase (MnP) and laccase [5, 7, 17, 20]. Nevertheless, not all of these enzymes are secreted by fungal cultures [7]. Lignin peroxidase (EC 1.11.1.14), also known as ligninase, is a heme-protein involves in oxidizing and/or cleaving of non-phenolic aromatic lignin moieties and similar molecules. Manganese peroxidase (EC 1.11.1.13) is a heme-containing glycoprotein, aids delignification by catalyzing reaction that oxidizes phenolic compounds in the presence of Mn^{2+}. Laccases (EC 1.10.3.2) are copper-containing oxidase enzymes that act on phenols and similar molecules by executing one-electron oxidations [5–7]. Versatile peroxidase (VP) is regarded as the third peroxidase, a LiP-MnP hybrid as it is capable of degrading both phenolic and non-phenolic lignin compounds and Mn^{2+} [6, 7].

2.3 Pretreatment of lignocelluloses with white-rot fungi

The sugarcane bagasse was subjected to fungal pretreatment by *P. ostreatus* and *C. subvermispora* for a period of 60 days [18]. At the end of pretreatment, *P. ostreatus* homogeneously degraded all the lignocellulose components of lignin, xylan and glucan up to 11.1, 15.7 and 8.4%, respectively. *C. subvermispora* yielded obvious lignin and xylan removal while consuming minimal glucan at 48, 47 and 13.6%, correspondingly. With sugarcane bagasse as the biomass, *P. ostreatus* behaves non-selectively due to the fact that the digestibility is not enhanced even when it degrades both lignin and polysaccharides. In contrast, *C. subvermispora* shows selective behavior as it removes lignin and xylan while sustaining glucan, which further improved the digestibility.

The biodegradability of wheat straw and oak wood chips treated with the white-rot fungi *C. subvermispora* and *L. edodes* was observed for 56 days [21]. Using wheat straw as the feedstock, *C. subvermispora* reached higher lignin, hemicellulose and cellulose degradation of 83.3, 80.5 and 20.2% than *L. edodes* with the values of 71.7, 69.3 and 12.2%, respectively. Different observation was found when choosing oak wood chips as the biomass. *C. subvermispora* achieved lower lignin, hemicellulose and cellulose removal of 53.5, 50.6 and 17.4% than *L. edodes* with the values of 60.6, 56.3 and 37.3%, correspondingly. Both fungi selectively degraded lignin in wheat straw and wood chips but with different strategy. *C. subvermispora* colonizes the biomass predominantly during the first 7 days and breaks lignin and hemicelluloses without growing, whereas *L. edodes* constantly grows and removes lignin during the growth. The relative lower lignin removal of wood chips compared to wheat straw indicates that the fungi had more difficulty to penetrate the wood chips due to its dense structure.

In a research done on pretreatment of willow sawdust via the white-rot fungi *A. biennis* and *Leiotrametes menziesii*, it was revealed that *A. biennis* was more preferable for fungal pretreatment even though it has lower delignification than *L. menziesii*, because it consumed a very low amount of cellulose [22]. After 30 days of treatment, the lignin, hemicellulose and cellulose loss attained by *A. biennis* were 17.1, 19.3 and 7.4%, respectively. On the other hand, higher lignin, hemicellulose and cellulose removal was achieved by *L. menziesii*, with the corresponding values of 30.5, 42.4 and 26.6%.

Xu et al. [23] reported that within 12 days of pretreatment, the highest lignin loss achieved by medicinal mushroom, *Inonotus obliquus*, using wheat straw as substrate is at 72%, with cellulose loss of 55%. However, lower delignification was observed for corn stover and rice straw of 47 and 39% with cellulose reduction of 55 and 45%. The hemicellulose content of wheat straw, corn stover and rice straw were decreased to 46, 39 and 44%, respectively. From these results, *I. obliquus* shows its potential as a delignifier of agricultural biomass as it can produce high-activity-level ligninolytic and hydrolytic enzymes.

T. versicolor and *S. hirsutum* showed selective delignification characteristics during the pretreatment of radiata pine wood chips [14]. Both fungi have the largest selectivity value on 21 days of treatment, with *T. versicolor* exhibited better selectivity than *S. hirsutum*. Both of *T. versicolor* and *S. hirsutum* delignified the chips by 16%. The hemicellulose and cellulose was reduced at 5% each for *T. versicolor* whereas 9% each for *S. hirsutum*. As the treatment period was increased, the selectivity values of both fungi decreases because cellulose was degraded together with lignin.

The delignification properties of two white-rot fungi, rainbow fungus (*T. versicolor*) and edible oyster fungus (*P. ostreatus*), on solid oriental beech wood (*Fagus orientalis* Lipsky) was studied for 120 days [10]. For both fungi, there is no substantial difference observed on lignin and cellulose degradation, with lignin degradation was more effective in the first 30 days of exposure. After 120 days of incubation, *T. versicolor* and *P. ostreatus* decayed lignin by 57.4 and 56.5%, and cellulose by 16.7 and 13.9%, respectively. Meanwhile, the decrease in total carbohydrate content was significantly higher for the first 30 days using *T. versicolor* as compared to *P. ostreatus*. At the end of the exposure period, the total carbohydrate content was almost the same, 7.3 and 6.7%, correspondingly. Both fungi are selective delignification, since the degradation of cellulose starts only after 60 days of incubation.

Four agricultural residues (wheat straw, corn stover, barley straw, and corncob) were pretreated for 21 days using the white-rot fungus *Irpex lacteus* [24]. The highest lignin removal was detected using corn stover (45.8%) as the feedstock, followed by wheat straw (42.3%), barley straw (31.0%) and corncob (17.1%). For glucan digestibility, the increment was significant for corn stover (up to 59.2%), wheat straw (up to 54.8%) and barley straw (up to 53.9%), except for corncob (reduced to 30.3%). The increase in xylan digestibility was observed in corn stover (up to 82.1%), wheat straw (up to 78.0%) and barley straw (up to 58.2%), but not for corncob (decreased to 22.4). Generally, all residues showed a reduction in lignin content. In the case of glucan and xylan digestibility, only corncob yielded lower digestibility after treatment. However, to be specific, *I. lacteus* behaves differently when subjected to different types of raw materials.

The lignin, hemicellulose and cellulose biodegradation of oil palm empty fruit bunches was investigated by exploiting two white-rot fungi, *P. ostreatus* and *P. chrysosporium* [25]. The lignin degradation was higher with *P. ostreatus* (51.9%) than with *P. chrysosporium* (42.1%) after treating for 21 days. In contrast, lower hemicellulose and cellulose degradation rates were noted for *P. ostreatus* (13.8 and 7.6%) compared to *P. chrysosporium* (27.7 and 28.2%). Since only a small amount of cellulose was degraded, fungal pretreatment using *P. ostreatus* is acceptable for palm residues. The fungus *P. ostreatus* can be considered as a selective delignifier because the cellulose degradation happens only after the 21 days of treatment, whereas *P. chrysosporium* is a non-selective delignifier as it concurrently breaks down lignin and structural carbohydrates.

Ishola et al. [26] found that fungal pretreatment improved the digestibility of oil palm empty fruit bunches by 4.5 times. The digestibility of untreated bunches was only 3.4%. This value was raised to 15.4% after the bunches were pretreated by *Pleurotus floridanus* fungus. After the pretreatment, the percentage of total lignin removal was very low, which is reduced by 0.03%. The hemicellulose content was increased by 4.4%, whereas the cellulose was decreased by 5.0% due to fungal attack on the linkage between lignin and carbohydrate.

Enhancement of hemicellulose accessibility was reported when fresh poplar wood (*Populus tomentosa*) was treated for 56 days with a common white-rot fungus on angiosperm wood, *Trametes velutina* [27]. Comparison between untreated and fungi-pretreated material revealed that lignin degradation can positively impact hemicellulose conversion. This was proven with the reduction in lignin content by 7.2% has resulted to an increase in both hemicellulose and cellulose content by 1.0 and 6.4%, consecutively. These findings suggested that lignin degradation rendered xylan more susceptible to xylanase and that in turn rendered cellulose more susceptible to cellulase.

For woody materials and agricultural residues feedstock, the ligninolytic systems and the appropriate fungal strains for the delignification may be different as they have different structure and chemical composition. Hence, it is important to discover the most significant white-rot fungal strain by assessing the strains for the highest degradation ability with the lowest holocellulose utilization as fungal selection subjects to the lignocellulosic biomass chosen for processing [11, 18]. Moreover, one fungus yields a very large difference of the decayed lignin-hemicellulose-cellulose ratio from another fungus, even when using different strains of the same species [6]. Some of recent researches on fungal pretreatment were tabulated in **Table 3**.

Substrate	Fungi species	Inoculum conc.	Moisture content (%)	T (°C)	pH	Time (days)	Nutrient	Effect	References
Sugarcane bagasse	*P. ostreatus*	0.05 w/w %	N.S	27	N.S	60	+	• Lignin, xylan and glucan degradation up to 11.1, 15.7 and 8.4% • Glucan and xylan digestibility reached 35 and 19%	[18]
	C. subvermispora							• Lignin, xylan and glucan removal at 48, 47 and 13.6% • Glucan and xylan digestibility increased up to 55 and 27%	
Wheat straw	*C. subvermispora*	10 w/w%	70%	24	N.S	56	N.S	• Decrease in lignin, hemicellulose and cellulose were 83.3, 80.5 and 20.2%	[21]
	L. edodes							• Lignin, hemicellulose and cellulose biodegradation of 71.7, 69.3 and 12.2%	
Oak wood chips	*C. subvermispora*	10 w/w%	70%	24	N.S	56	N.S	• Reduction of lignin, hemicellulose and cellulose content at 53.5, 50.6 and 17.4%	[21]
	L. edodes							• Lignin, hemicellulose and cellulose removal increased to 60.6, 56.3 and 37.3%	
Willow sawdust	*A. biennis*	0.32 w/w %	80%	27	N.S	30	—	• Degradation of lignin, hemicellulose and cellulose reached 17.1, 19.3 and 7.4%	[22]
	L. menziesii	0.48 w/w %						• Lignin, hemicellulose and cellulose loss of 30.5, 42.4 and 26.6%	
Wheat straw	*I. obliquus*	8%	N.S	28	6	12	+	• Decrease in lignin, hemicellulose and cellulose up to 72, 46 and 55%	[23]
Corn stover									

Substrate	Fungi species	Inoculum conc.	Moisture content (%)	T (°C)	pH	Time (days)	Nutrient	Effect	References
								• Lignin, hemicellulose and cellulose reduction reached 47, 39 and 55%	
Rice straw								• Removal of lignin, hemicellulose and cellulose increased up to 39, 44 and 45%	
Radiata pine	*T. versicolor*	N.S	70%	25	N.S	21	N.S	• Loss of lignin at 16%, while both hemicellulose and cellulose at 5% each	[14]
	S. hirsutum							• Lignin degradation of 16%, whereas hemicellulose and cellulose of 9% individually	
Beech wood	*T. versicolor*	N.S	65%	22	N.S	120	N.S	• Reduction of lignin and cellulose up to 57.4 and 16.7%	[10]
	P. ostreatus							• Lignin and cellulose biodegradation of 56.5 and 13.9%	
Corn stover	*I. lacteus*	50 v/w%	7.3–8.5%	30	N.S	21	N.S	• Removal of lignin reached 45.8% • Glucan and xylan digestibility increased up to 59.2 and 82.1%	[24]
Wheat straw								• Lignin loss was 42.3% • Digestibility of glucan and xylan reached 54.8 and 78.0%	
Barley straw								• Lignin removal up to 31.0% • Digestibility of glucan and xylan enhanced to 53.9 and 58.2%	
Corncob								• Degradation of lignin at 17.1% • Glucan and xylan digestibility reduced to 30.3 and 22.4%	

Substrate	Fungi species	Inoculum conc.	Moisture content (%)	T (°C)	pH	Time (days)	Nutrient	Effect	References
Oil palm empty fruit bunches	*P. ostreatus*	N.S	67%	30	N.S	21	N.S	• Lignin, hemicellulose and cellulose degradation at 51.9, 13.8 and 7.6%	[25]
	P. chrysosporium							• Reduction of lignin, hemicellulose and cellulose were 42.1, 27.7 and 28.2%	
Oil palm empty fruit bunches	*P. floridanus*	N.S	59.40%	31	N.S	28	+	• Lignin and cellulose were reduced by 0.03 and 5.0%, while hemicellulose was increased by 4.4% • Digestibility was improved by 4.5 times	[26]
Poplar wood	*T. velutina*	100 v/w%	N.S	28	N.S	56	N.S	• Delignification by 7.2%, whereas both hemicellulose and cellulose were increased by 1.0 and 6.4%	[27]

N.S, not specified.

Table 3.
Summary of recent publications on fungal pretreatment.

3. Parameters affecting pretreatment process

High lignin degradation can be achieved by having high activities of white-rot fungi and production of ligninolytic enzymes. This is influenced by several pretreatment parameters such as fungal strain, inoculum concentration, moisture content, aeration, pH, temperature, supplements and incubation time [28–30]. Moreno et al. [29] reviewed that for solid state fermentations (SSF), depending on the strain used, the usual conditions that have been used are at moisture content 45–85%, pH 4–5, with an inoculum level of 1–10 mg/g substrate (dry weight), at temperatures ranging from 15 to 40°C and over 1–12 weeks. The optimization of these parameters is important to increase the efficiency of the pretreatment by reducing the carbohydrate loss and pretreatment time [31]. However, for most of these factors, the optimal conditions are depended on the substrate and fungal strain [28]. Temperature and pH are reported to affect fungal metabolism, spore germination and growth. Low moisture content can reduce nutrients availability and growth, while higher moisture content can boost contamination, reduce heat and oxygen transmission, and affect enzyme production [32]. Adekunle et al. [33] reported that the pH and temperature of the SSF play a vital role in the production of laccase by *T. versicolor*.

3.1 Fungal strain

In order for the white-rot fungi to be used in the pretreatment process, screening of a large number of fungal isolates is important in order to have the right isolates for the process. Screening step allows the selection of isolates with the highest ligninolytic enzymes production and activity as well as high lignin degradation on the specific substrates. In order to limit matter losses, selective delignification is crucial and high fermentable sugar losses must be avoided. White-rot fungi strains should therefore be carefully selected based on these important parameters. **Table 4** summarizes some of white-rot fungi species that have been studied for several lignocellulosic biomass pretreatment.

3.2 Inoculum concentration

Inoculum concentration is an important factor in biological pretreatment. Sufficient amounts of inoculum must be defined to ensure good fungal growth and substrate colonization. The time required for the colonization of the substrate is affected by the type and amount of inoculums [31]. Higher concentration of inoculum will lead to shorter time of colonization of the substrate [49].

3.3 Moisture content

Moisture content of the solid state fermentation is a critical aspect for fungal growth and activities. Lignin degradation is significantly influenced by this factor as it affects the growth and activities of the fungal. Increasing the moisture content enhances the nutrient transfer but reduces the porosity of the substrate and limits oxygen transfer [28]. However, insufficient water content in the substrates may cause deactivation of the fungi. Optimum moisture content depends upon the organism and the substrate used for SSF [30]. The range of moisture content of substrate for SSF using white-rot fungi is usually between 45 and 85% [29]. A study on the effect of moisture content for delignification of cotton stalks by *Daedalea flavida* MTCC 145 (DF-2) in SSF found that the highest ligninolytic enzyme

Fungi species	Substrate(s)	Enzymes	References
Ceriporiopsis subvermispora	Hazel branches	Laccase and MnP	[34]
	Albizia chips		[35]
	Miscanthus sinensis		[36]
Echinodontium taxodii	Bamboo	Laccase and MnP	[37]
Trametes versicolor	Pine wood chips	Laccase	[14]
	Corn stalk		[33]
	Corn silage		[38]
Tricholoma giganteum	Wheat straw	Laccase	[39]
Schizophyllum commune	Banana stalk Corn cobs Sugarcane bagasse Wheat straw	Laccase, LiP and MnP	[40, 41]
Pseudolagarobasidium acaciicola	*Parthenium* biomass	Laccase	[42]
Dichomitus squalens	Chestnut shell	Laccase	[43]
Daedalea flavida	Cotton stalks	Laccase and LiP	[44]
Tremetes villosa	Coconut shell Sugarcane bagasse Sisal fiber	MnP	[45]
Stereum ostrea	Wheat bran	MnP	[46]
Pleurotus ostreatus	Rice straw	N.S	[47]
Polyporus brumalis	Wheat straw	N.S	[48]

N.S,not specified.

Table 4.
Lignocellulosic biomass pretreatments with different white-rot fungi species and their isolated enzymes.

activities, optimal lignin degradation 29.88 $\pm$ 0.97% (w/w) with cellulose loss 11.70 $\pm$ 1.30% (w/w) were observed at 75% moisture content [44]. It was reported that the lignin degradation increased with increase in moisture content. Cellulose and hemicellulose degradation were found to be increased at higher moisture content and small particle size. The selectivity value, SV also influenced by the moisture content. Increase in moisture content decreased the SV, and this may be due to the decreasing of lignin degradation compared to cellulose loss caused by oxygen diffusion declining and ligninolytic enzymes inhibition. Similar condition was also reported for SSF of steam-exploded cornstalk by *T. versicolor* where the highest laccase activity achieved in this study was 2765.81 Ug^{-1} at 75% moisture content [33]. A study on laccase production by a novel white-rot fungus *Pseudolagarobasidium acaciicola* LA 1 through SSF of *Parthenium* biomass reported that the highest laccase activities, 16,388 Ug^{-1} of substrate was found at liquid to solid ratio of 5 with an incubation period of 7 days [42].

3.4 Temperature

Temperature is another very critical factor in the pretreatment using white-rot fungi. However, different genus has different tolerant to temperature. Fungal physiology, fungal strain and types of substrate also resulted in different optimal temperature for biological pretreatment [30]. This statement is in agreement with several studies which showed the production of ligninolytic enzyme using white-rot fungi has various optimal temperature [33, 50]. White-rot basidiomycetes grow optimally at temperature between 25 and 30°C while most of white-rot ascomycetes fungi grow optimally at 39°C [5]. The metabolism of these fungi produces heat and develops temperature gradients in SSF media. The accumulated heat can lead to adverse effect on the fungal growth and their metabolic activity which leading to the denaturation of the key enzymes. From the studies on pretreatment of rice straw, ligninolytic activities by *S. commune* was found to be peaked at 30 and 35°C [40, 41]. Meanwhile the highest ligninases production by *T. versicolor* was reported at 40°C [51]. Adekunle et al. [33] reported in their study on the SSF of steam-exploded cornstalk with *T. versicolor* that there was a direct correlation between the temperature and laccase production, with the highest laccase activity of 2677.16 U g^{-1} was produced at 28°C. The maximum production of laccase by *T. giganteum* AGHP (1.53×10^5 Ug^{-1} of dry substrate) was obtained at 30°C [39]. This study also showed that lower temperature of 10 and 20°C are not suitable for the growth of fungi due to lower enzyme production. Similar result was obtained from the study on laccase production by *Pseudolagarobasidium acaciicola* LA 1, the optimum production (19,944 Ug^{-1} dry substrate), was found at 30°C [42].

3.5 pH

pH is one of the prominent parameters in the cultivation of fungi and it is very problematic to control in SSF [52]. The initial pH of the medium influences the microbial growth and the production of ligninolytic enzyme. White-rot fungi grow well at pH 4–5, while substrate acidity decreases their growth. A study conducted on the isolation of laccase from a novel white-rot fungus *Pseudolagarobasidium acaciicola* LA 1 through SSF of *Parthenium* biomass showed that the isolated laccase was found to perform optimally at pH 4.5 and highly stable within the range of pH 4–7 for 24 h [42]. The effect of pH is important in the case of laccase production, and a small change in intracellular pH will result in a decrease of macromolecules synthesis. Patel and Gupte [39] reported that the maximal laccase production (1.27×10^5 Ug^{-1} of dry substrate) by *Tricholoma giganteum* AGHP was achieved at pH 5.0. No increment in the production of enzyme was found at higher pH. This may be attributed to the poor mycelial growth at an elevated pH which may restrict the laccase production. It was reported that the maximum ligninolytic activities by *T. versicolor* were found at pH 4.0 and 5.0 [33]. Asgher et al. [40] showed that the optimum enzymes production by *S. commune* IBL-06 was found to be at pH 5 while pH 4.43 and 4.46 for *S. commune* NI-07 [41]. Change in pH will affect the three dimensional structure of laccase which in turn leads to the decrease in laccase activity [5].

3.6 Aeration

Production and activity of ligninolytic enzymes are also influenced by aeration. There are several purposes of aeration such as to supply oxygen into the media, for the removal of CO_2, heat dissipation, distribution of water vapor to regulate humidity, and circulation of volatile compounds produced during metabolism. Thus, this factor should be optimized to improve rate of delignification [49].

3.7 Supplements

Other factor such as various supplements (Cu^{2+}, Mn^{2+}, ferulic acid, xylidine, veratric acid, vanillic acid, cinnamic acid, guaiacol, etc.) for the SSF media have previously been reported in studying their effect on production of ligninolytic enzymes [44, 53]. Many studies reported that the copper at various concentrations influences laccase production in *S. ostrea*, *T. pubescens*, *P. eryngii*, and *P. ostreatus* [46, 54]. This is related to the role of Cu^{2+} that controls the transcription of laccase gene and also enhances the stability of this enzyme. Meanwhile, the concentration on Mn^{2+} influenced both MnP and laccase production by different *Pleurotus* species [55].

4. Conclusion

Pretreatment is very crucial in the conversion of lignocellulosic materials to other value-added products as lignin acts as the barrier for enzyme penetration. Comparing various pretreatment strategies, fungal pretreatment is more favorable because it is an environmental-friendly process. White-rot fungi with high selectivity of delignification than cellulose removal are more desirable compared to other microorganisms as cellulose is the feedstock for the subsequent hydrolysis process. Fungal strain, inoculum concentration, moisture content, temperature, pH, aeration and supplements are crucial parameters for fungal growth and metabolism to achieve good pretreatment outcome.

Author details

Najiah Nadir, Nur Liyana Ismail and Azlan Shah Hussain*
PETRONAS Research Sdn. Bhd., Selangor, Malaysia

*Address all correspondence to: azlanshah_hh@petronas.com.my

References

[1] Lee RA, Lavoie JM. From first- to third-generation biofuels: Challenges of producing a commodity from a biomass of increasing complexity. Animal Frontiers. 2013;**3**(2):6-11. DOI: 10.2527/af.2013-0010

[2] Haghighi Mood S, Hossein Golfeshan A, Tabatabaei M, Salehi Jouzani G, Najafi GH, Gholami M, et al. Lignocellulosic biomass to bioethanol, a comprehensive review with a focus on pretreatment. Renewable and Sustainable Energy Reviews. 2013;**27**: 77-93. DOI: 10.1016/j.rser.2013.06.033

[3] Zheng Y, Zhao J, Xu F, Li Y. Pretreatment of lignocellulosic biomass for enhanced biogas production. Progress in Energy and Combustion Science. 2014;**42**:35-53. DOI: 10.1016/j.pecs.2014.01.001

[4] Ravindran R, Jaiswal AK. A comprehensive review on pre-treatment strategy for lignocellulosic food industry waste: Challenges and opportunities. Bioresource Technology. 2016;**199**: 92-102. DOI: 10.1016/j.biortech.2015.07.106

[5] Sindhu R, Binod P, Pandey A. Biological pretreatment of lignocellulosic biomass—An overview. Bioresource Technology. 2016;**199**: 76-82. DOI: 10.1016/j.biortech.2015.08.030

[6] Narayanaswamy N, Dheeran P, Verma S, Kumar S. Biological pretreatment of lignocellulosic biomass for enzymatic saccharification. In: Fang Z, editor. Pretreatment Techniques for Biofuels and Biorefineries. Green Energy. Berlin: Springer; 2013. pp. 3-34. DOI: 10.1007/978-3-642-32735-3_1

[7] Wan C, Li Y. Fungal pretreatment of lignocellulosic biomass. Biotechnology Advances. 2012;**30**(6):1447-1457. DOI: 10.1016/j.biotechadv.2012.03.003

[8] Taha M, Shahsavari E, Al-Hothaly K, Mouradov A, Smith AT, Ball AS, et al. Enhanced biological straw saccharification through coculturing of lignocellulose-degrading microorganisms. Applied Biochemistry and Biotechnology. 2015;**175**(8): 3709-3728. DOI: 10.1007/s12010-015-1539-9

[9] Chaturvedi V, Verma P. An overview of key pretreatment processes employed for bioconversion of lignocellulosic biomass into biofuels and value added products. 3 Biotech. 2013;**3**(5):415-431. DOI: 10.1007/s13205-013-0167-8

[10] Bari E, Taghiyari HR, Naji HR, Schmidt O, Ohno KM, Clausen CA, et al. Assessing the destructive behaviors of two white-rot fungi on beech wood. International Biodeterioration & Biodegradation. 2016;**114**:129-140. DOI: 10.1016/j.ibiod.2016.06.010

[11] Saha BC, Qureshi N, Kennedy GJ, Cotta MA. Biological pretreatment of corn Stover with white-rot fungus for improved enzymatic hydrolysis. International Biodeterioration & Biodegradation. 2016;**109**:29-35. DOI: 10.1016/j.ibiod.2015.12.020

[12] Hermosilla E, Rubilar O, Schalchli H, da Silva AS, Ferreira-Leitao V, Diez MC. Sequential white-rot and brown-rot fungal pretreatment of wheat straw as a promising alternative for complementary mild treatments. Waste Management. 2018;**79**:240-250. DOI: 10.1016/j.wasman.2018.07.044

[13] Xu G, Wang L, Liu J, Wu J. FTIR and XPS analysis of the changes in bamboo chemical structure decayed by white-rot and brown-rot fungi. Applied Surface Science. 2013;**280**:799-805. DOI: 10.1016/j.apsusc.2013.05.065

[14] Shirkavand E, Baroutian S, Gapes DJ, Young BR. Pretreatment of radiata

pine using two white rot fungal strains *Stereum hirsutum* and *Trametes versicolor*. Energy Conversion and Management. 2017;**142**:13-19. DOI: 10.1016/j.enconman.2017.03.021

[15] Monrroy M, Ibañez J, Melin V, Baeza J, Mendonça RT, Contreras D, et al. Bioorganosolv pretreatments of *P. radiata* by a brown rot fungus (*Gloephyllum trabeum*) and ethanolysis. Enzyme and Microbial Technology. 2010;**47**(1–2):11-16. DOI: 10.1016/j.enzmictec.2010.01.009

[16] Nilsson T, Daniel G. Chemistry and microscopy of wood decay by some higher ascomycetes. Holzforschung. 1989;**43**(1):11-18. DOI: 10.1515/hfsg.1989.43.1.11

[17] Castoldi R, Bracht A, de Morais GR, Baesso ML, Correa RCG, Peralta RA, et al. Biological pretreatment of Eucalyptus grandis sawdust with white-rot fungi: Study of degradation patterns and saccharification kinetics. Chemical Engineering Journal. 2014;**258**:240-246. DOI: 10.1016/j.cej.2014.07.090

[18] da Silva Machado A, Ferraz A. Biological pretreatment of sugarcane bagasse with basidiomycetes producing varied patterns of biodegradation. Bioresource Technology. 2017;**225**:17-22. DOI: 10.1016/j.biortech.2016.11.053

[19] Ginns JH. Polypores of British Columbia (Fungi: Basidiomycota) [Internet]. Tech Rep 1. Vol. 104. Victoria, BC: Province of British Columbia; 2017. 1-260 p. Available from: https://www.for.gov.bc.ca/hfd/pubs/Docs/Tr/TR104.pdf [Accessed: 2018-12-28]

[20] Wang FQ, Xie H, Chen W, Wang ET, Du FG, Song AD. Biological pretreatment of corn Stover with ligninolytic enzyme for high efficient enzymatic hydrolysis. Bioresource Technology. 2013;**144**:572-578. DOI: 10.1016/j.biortech.2013.07.012

[21] van Kuijk SJA, Sonnenberg ASM, Baars JJP, Hendriks WH, del Río JC, Rencoret J, et al. Chemical changes and increased degradability of wheat straw and oak wood chips treated with the white rot fungi *Ceriporiopsis subvermispora* and *Lentinula edodes*. Biomass and Bioenergy. 2017;**105**:381-391. DOI: 10.1016/j.biombioe.2017.07.003

[22] Alexandropoulou M, Antonopoulou G, Fragkou E, Ntaikou I, Lyberatos G. Fungal pretreatment of willow sawdust and its combination with alkaline treatment for enhancing biogas production. Journal of Environmental Management. 2017;**203**:704-713. DOI: 10.1016/j.jenvman.2016.04.006

[23] Xu X, Xu Z, Shi S, Lin M. Lignocellulose degradation patterns, structural changes, and enzyme secretion by *Inonotus obliquus* on straw biomass under submerged fermentation. Bioresource Technology. 2017;**241**:415-423. DOI: 10.1016/j.biortech.2017.05.087

[24] García-Torreiro M, López-Abelairas M, Lu-Chau TA, Lema JM. Fungal pretreatment of agricultural residues for bioethanol production. Industrial Crops and Products. 2016;**89**:486-492. DOI: 10.1016/j.indcrop.2016.05.036

[25] Piñeros-Castro Y, Velásquez-Lozano M. Biodegradation kinetics of oil palm empty fruit bunches by white rot fungi. International Biodeterioration & Biodegradation. 2014;**91**:24-28. DOI: 10.1016/j.ibiod.2014.03.009

[26] Ishola MM, Isroi TMJ. Effect of fungal and phosphoric acid pretreatment on ethanol production from oil palm empty fruit bunches (OPEFB). Bioresource Technology. 2014;**165**:9-12. DOI: 10.1016/j.biortech.2014.02.053

[27] Wang W, Yuan T, Cui B, Dai Y. Investigating lignin and hemicellulose in white rot fungus-pretreated wood that

affect enzymatic hydrolysis. Bioresource Technology. 2013;**134**:381-385. DOI: 10.1016/j.biortech.2013.02.042

[28] Rouches E, Herpoël-Gimbert I, Steyer JP, Carrere H. Improvement of anaerobic degradation by white-rot fungi pretreatment of lignocellulosic biomass: A review. Renewable and Sustainable Energy Reviews. 2016;**59**: 179-198. DOI: 10.1016/j.rser.2015.12.317

[29] Moreno AD, Ibarra D, Alvira P, Tomás-Pejó E, Ballesteros M. A review of biological delignification and detoxification methods for lignocellulosic bioethanol production. Critical Reviews in Biotechnology. 2015; **35**(3):342-354. DOI: 10.3109/ 07388551.2013.878896

[30] Isroi MR, Syamsiah S, Niklasson C, Cahyanto MN, Ludquist K, et al. Biological pretreatment of lignocelluloses with white-rot fungi and its applications: A review. BioResources. 2011;**6**(4):5224-5259. DOI: 10.15376/ biores.6.4.5224-5259

[31] van Kuijk SJA, Sonnenberg ASM, Baars JJP, Hendriks WH, Cone JW. Fungal treated lignocellulosic biomass as ruminant feed ingredient: A review. Biotechnology Advances. 2015;**33**(1): 191-202. DOI: 10.1016/j.biotechadv. 2014.10.014

[32] Chen H. Biotechnology principles of solid state fermentation. In: Modern Solid State Fermentation. Dordrecht: Springer; 2013. pp. 23-74. DOI: 10.1007/ 978-94-007-6043-1_2

[33] Adekunle AE, Zhang C, Guo C, Liu CZ. Laccase production from *Trametes versicolor* in solid-state fermentation of steam-exploded pretreated cornstalk. Waste and Biomass Valorization. 2017; **8**(1):153-159. DOI: 10.1007/s12649-016-9562-9

[34] Liu X, Hiligsmann S, Gourdon R, Bayard R. Anaerobic digestion of lignocellulosic biomasses pretreated with *Ceriporiopsis subvermispora*. Journal of Environmental Management. 2017; **193**:154-162. DOI: 10.1016/j. jenvman.2017.01.075

[35] Ge X, Matsumoto T, Keith L, Li Y. Fungal pretreatment of albizia chips for enhanced biogas production by solid-state anaerobic digestion. Energy & Fuels. 2015;**29**(1):200-204. DOI: 10.1021/ef501922t

[36] Vasco-Correa J, Li Y. Solid-state anaerobic digestion of fungal pretreated *Miscanthus sinensis* harvested in two different seasons. Bioresource Technology. 2015;**185**:211-217. DOI: 10.1016/j.biortech.2015.02.099

[37] Kong W, Chen H, Lyu S, Ma F, Yu H, Zhang X. Characterization of a novel manganese peroxidase from white-rot fungus *Echinodontium taxodii* 2538, and its use for the degradation of lignin-related compounds. Process Biochemistry. 2016;**51**(11):1776-1783. DOI: 10.1016/j.procbio.2016.01.007

[38] Tišma M, Planinić M, Bucić-Kojić A, Panjičko M, Zupančič GD, Zelić B. Corn silage fungal-based solid-state pretreatment for enhanced biogas production in anaerobic co-digestion with cow manure. Bioresource Technology. 2018;**253**:220-226. DOI: 10.1016/j.biortech.2018.01.037

[39] Patel H, Gupte A. Optimization of different culture conditions for enhanced laccase production and its purification from *Tricholoma giganteum* AGHP. Bioresources and Bioprocessing. 2016;**3**(1):11-20. DOI: 10.1186/ s40643-016-0088-6

[40] Asgher M, Wahab A, Bilal M, Nasir Iqbal HM. Lignocellulose degradation and production of lignin modifying enzymes by *Schizophyllum commune* IBL-06 in solid-state fermentation. Biocatalysis and Agricultural

Biotechnology. 2016;**6**:195-201. DOI: 10.1016/j.bcab.2016.04.003

[41] Kumar VP, Naik C, Sridhar M. Production, purification and characterization of novel laccase produced by *Schizophyllum commune* NI-07 with potential for delignification of crop residues. Applied Biochemistry and Microbiology. 2015;**51**(4):432-441. DOI: 10.1134/S0003683815040080

[42] Adak A, Tiwari R, Singh S, Sharma S, Nain L. Laccase production by a novel white-rot fungus *Pseudolagarobasidium acaciicola* LA 1 through solid-state fermentation of *Parthenium* biomass and its application in dyes decolorization. Waste and Biomass Valorization. 2016;**7**(6):1427-1435. DOI: 10.1007/s12649-016-9550-0

[43] Zhang X, Yu H, Huang H, Liu Y. Evaluation of biological pretreatment with white rot fungi for the enzymatic hydrolysis of bamboo culms. International Biodeterioration & Biodegradation. 2007;**60**(3):159-164. DOI: 10.1016/j.ibiod.2007.02.003

[44] Meehnian H, Jana AK, Jana MM. Effect of particle size, moisture content, and supplements on selective pretreatment of cotton stalks by *Daedalea flavida* and enzymatic saccharification. 3 Biotech. 2016;**6**(2): 235. DOI: 10.1007/s13205-016-0548-x

[45] Silva MLC, de Souza VB, da Silva Santos V, Kamida HM, de Vasconcellos-Neto JRT, Góes-Neto A, et al. Production of manganese peroxidase by *Trametes villosa* on unexpensive substrate and its application in the removal of lignin from agricultural wastes. Advances in Bioscience and Biotechnology. 2014;**5**:1067-1077. DOI: 10.4236/abb.2014.514122

[46] Usha KY, Praveen K, Reddy BR. Enhanced production of ligninolytic enzymes by a mushroom *Stereum ostrea*. Biotechnology Research International. 2014;**2014**:1-9. DOI: 10.1155/2014/815495

[47] Mustafa AM, Poulsen TG, Sheng K. Fungal pretreatment of rice straw with *Pleurotus ostreatus* and *Trichoderma reesei* to enhance methane production under solid-state anaerobic digestion. Applied Energy. 2016;**180**:661-671. DOI: 10.1016/j.apenergy.2016.07.135

[48] Rouches E, Zhou S, Sergent M, Raouche S, Carrere H. Influence of white-rot fungus *Polyporus brumalis* BRFM 985 culture conditions on the pretreatment efficiency for anaerobic digestion of wheat straw. Biomass & Bioenergy. 2018;**110**:75-79. DOI: 10.1016/j.biombioe.2018.01.018

[49] Madadi M, Abbas A. Lignin degradation by fungal pretreatment: A review. Journal of Plant Pathology & Microbiology. 2017;**8**(2):1-6. DOI: 10.4172/2157-7471.1000398

[50] Tian X, Fang Z, Guo F. Impact and prospective of fungal pre-treatment of lignocellulosic biomass for enzymatic hydrolysis. Biofuels, Bioproducts and Biorefining. 2012;**6**(3):335-350. DOI: 10.1002/bbb.346

[51] Asgher M, Iqbal HMN, Asad MJ. Kinetic characterization of purified laccase produced from *Trametes versicolor* IBL-04 in solid state bio-processing of corncobs. BioResources. 2012;**7**(1):1171-1188. DOI: 10.15376/biores.7.1.1171-1188

[52] Hölker U, Höfer M, Lenz J. Biotechnological advantages of laboratory-scale solid-state fermentation with fungi. Applied Microbiology and Biotechnology. 2004; **64**(2):175-186. DOI: 10.1007/s00253-003-1504-3

[53] Liu Y, Juan L, Luo Z, Rao S, Su Y, Yang Y. Effect of supplements Mn^{2+}, Cu^{2+}, and aromatic compounds and

Penicillium decumbens on lignocellulosic enzyme activity and productivity of *Catathelasma ventricosum*. Journal of Microbiology and Biotechnology. 2013; **23**:565-571. DOI: 10.4014/jmb.1211.11007

[54] Galhaup C, Goller S, Peterbauer CK, Strauss J, Haltrich D. Characterization of the major laccase isoenzyme from *Trametes pubescens* and regulation of its synthesis by metal ions. Microbiology. 2002;**148**(7):2159-2169. DOI: 10.1099/00221287-148-7-2159

[55] Stajić M, Persky L, Hadar Y, Friesem D, Duletić-Laušević S, Wasser SP, et al. Effect of copper and manganese ions on activities of laccase and peroxidases in three *Pleurotus* species grown on agricultural wastes. Applied Biochemistry and Biotechnology. 2006; **128**(1):87-96. DOI: 10.1385/ABAB:128:1:087

Chapter 4

Woody Feedstock Pretreatments to Enhance Pyrolysis Bio-oil Quality and Produce Transportation Fuel

Hamid Rezaei, Fahimeh Yazdanpanah, Jim Choon Lim, Anthony Lau and Shahab Sokhansanj

Abstract

Lignocellulosic biomass as a potential renewable source of energy has a near-zero CO_2 emission. Pyrolysis converts biomass to a liquid fuel and increases the energy density and transportability. The pyrolysis bio-oil shows promising properties to substitute the conventional fossil fuels. But, unprocessed biomass is low in bulk and energy density; high in moisture; heterogeneous in physical and chemical properties, highly hygroscopic and difficult to handle. That is why the biomass needs mechanical, chemical and/or thermal pretreatments to turn into a more homogeneous feedstock and minimize the post-treatment fuel upgrading. This chapter explains the effects that various pretreatments such as size reduction, drying, washing and thermal pretreatments have on the quality and quantity of bio-oil. Washing with water or acid/alkali solutions extracts the minerals that consequently reduces the ash and shortens the reactor clean-out cycle. Torrefaction is gaining attention as an effective pretreatment to modify the quality of biomass in terms of physical and chemical properties. Torrefaction produces a uniform biomass with lower moisture, acidity and oxygen contents and higher energy density and grindability than raw biomass. Pyrolysis of torrefied biomass produces bio-oil with enhanced compositional and physical properties such as a higher heating value and increased C (lower O/C ratio).

Keywords: pyrolysis, pretreatment, torrefaction, washing, bio-oil, bio-fuel

1. Introduction

Crude oil and coal with high energy density and stable properties to store and transport are preferred fuels for industrial and home applications [1]. The recent reduction in affordable fossil fuels and environmental concerns increased greenhouse gas (GHG) emissions from these fuels that have motivated the public and private sectors to develop renewable energy sources. In recent years, the governments of Canada and the USA adopted regulations to phase out coal-fired power plants by 2030 and investigate other resources to reduce GHG emission. Biomass feedstocks are considered as the renewable resource with a near-zero CO_2 input-output balance [2–4]. Biomass is the only renewable energy source which may be used in all solid, liquid and gas phases [5]. But compared to fossil fuels, biomass is lower in energy (17–19 MJ/kg) and bulk density (60–100 kg/m^3) [6];

heterogeneous in physical, chemical and thermal properties; high in moisture [6], mineral [7] and oxygen contents [8]; highly hygroscopic [9] and difficult to handle [6]. Converting biomass to secondary liquid or gaseous fuels through thermal conversion processes is a way of increasing the energy density and transportability. Mechanical, chemical and thermal pretreatments are able to modify the biomass properties in order to produce a more homogeneous fuel and minimize the post-treatment fuel upgrading.

The current article reviews the feedstock properties that are important for the fast pyrolysis process and the potential pretreatments to modify the biomass properties are explained. Specifically, drying, grinding, washing and torrefaction processes and their influence on the final bio-oil product are thoroughly explained.

2. Pyrolysis

Pyrolysis is a thermal process to convert biomass particles to secondary solid, liquid and/or gaseous fuels [10]. The liquid bio-oil obtained from pyrolysis of biomass shows promising properties to substitute the conventional fossil fuels. The breakdown of biomass particles takes place in a pyrolysis reactor by exposing them to an oxygen-free or deficit environment. As the temperature of the particle increases, different stages of thermal treatment occur. Drying happens first when a moist particle is heated up to temperatures of about 150°C [11]. At temperatures of 200–300°C and in the absence of oxygen/air, various degrees of torrefaction happen and release volatile [12–14]. At temperatures higher than 300°C, more severe decomposition of biomass happens and structural changes such as a change in chemical composition and porosity of particles occur in the solid matrix.

Pyrolysis processes are generally categorized as "slow" and "fast" according to the time taken for processing the feed into pyrolysis products. Slow pyrolysis utilizes low temperatures of 300–400°C over a residence time of 30 minutes to hours to maximize char formation. Fast pyrolysis is a rapid thermal decomposition process to maximize the liquid fraction of products. In industrial fast pyrolysis, biomass particles would be exposed suddenly to an environment of 400–600°C to heat up rapidly. The combined effects of moisture content, particle size [15–17] and reactor temperature [18, 19] make a non-linear temperature profile inside the particle which affects the rate of conversion. The residence time of the volatile fraction in the pyrolysis reactor and its contact with solid particles should be less than 2–3 seconds to prevent the secondary reactions. Pyrolysis takes place in a complete absence of oxygen. The volatiles are quickly removed and quenched to maximize liquid yield. **Figure 1** displays a schematic diagram of an industrial pyrolysis process. In the following sections, a comprehensive literature review is presented to discuss the effects of various parameters related to feedstock and pre-treatments on the efficiency of the pyrolysis process.

2.1 Feedstock composition

A plant-based dry biomass has the main fractions of cellulose, hemicelluloses, lignin, extractives and minerals. Wood extractives, or wood extracts, are low molecular weight molecules that are extracted from wood using solvents or other extraction methods. The extractives are the waxes, fatty acids, resin acids and terpenes of a tree. **Table 1** lists the composition of a broad range of biomass species. The composition of biomass changes the chemical composition and heating value of pyrolysis products [33]. Higher lignin content reduces the bio-oil yield at

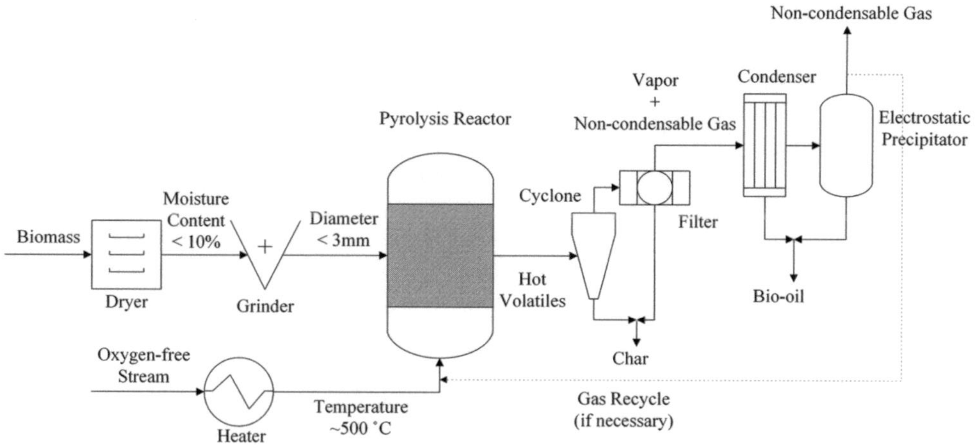

Figure 1.
Schematic diagram of the pyrolysis process, including drying, grinding, pyrolyzer, separation stages and condenser.

Biomass species	Composition (wt.%)			Refs.	Ultimate analysis (wt.%)			Ash (wt.%)	HHV (MJ/kg)	Refs.
	Cel.[1]	Hem.[2]	Lig.[3]		C	H	O			
Pine wood	42.1	27.7	25.0	[20]	45.9	5.3	48.2	0.35	18.98	[21]
Beech wood	41.7	37.1	18.9	[22]	46.7	5.7	47.6	1.04	–	[22]
Douglas fir	50.0	17.8	28.3	[23]	50.7	6.0	42.5	0.50	20.07	[24]
Spruce wood	41.1	20.9	28.0	[20]	48.9	6.0	44.6	0.30	19.26	[20]
Poplar aspen	49.9	22.4	18.1	[25]	–	–	–	–	–	–
Poplar wood	42.2	16.6	25.6	[26]	48.4	5.8	43.7	1.43	19.71	[26]
Birch wood	35.7	25.1	19.3	[20]	49.0	6.3	44.1	0.30	18.40	[27]
Corn stover	36.4	22.6	16.6	[28]	–	–	–	–	–	–
Corn cob	52.0	32.5	15.5	[29]	49.0	5.6	43.8	1.10	–	[29]
Wheat straw	38.2	24.7	23.4	[28]	40.7	5.8	52.9	10.58	16.24	[30]
Rice straw	34.2	24.5	23.4	[31]	36.9	5.0	37.9	11.70	16.78	[31]
Rice straw	32.1	26.5	12.5	[32]						
Olive husk	25.2	24.2	50.6	[29]	50.2	6.4	38.4	4.10	–	[29]
Tea waste	33.2	23.3	43.5	[29]	48.2	5.5	44.3	1.50	–	[29]

[1]*Cellulose.*
[2]*Hemicellulose.*
[3]*Lignin.*

Table 1.
Composition and ultimate analysis, ash content and calorific value of various biomass feedstock.

the expense of higher biochar yield in the pyrolysis process [29]. The extractives have the highest heating value in the woody biomass. The heating values of biomass components are 17–18, 16–17, 25–26 and 33–38 GJ/tonne for cellulose, hemicellulose, lignin and extractives, respectively [34].

Each sub-component of the biomass has a specific thermal reactivity. For example, cellulose decomposes at 300–400°C; hemicellulose decomposes at 200–300°C and lignin decompose continuously in the wide temperature range of

180–600°C [35]. This specification directly influences the temperature range at which the material thermally decomposes and consequently the optimum pyrolysis temperature at which the maximum bio-oil yield is obtained. **Table 2** lists pyrolysis liquid yield of various biomass species at tested temperatures and depicts the variability of produced bio-oil yield among the pyrolysis of different feedstocks. Apart from variability in conversion rate among species, the conversion rate increases with the reaction temperature.

In addition to the chemical composition, the elemental and proximate analysis of the biomass feedstock changes the properties of the produced fuel. From the elemental point of view, the biomass is mainly contained of carbon, hydrogen and oxygen (**Table 1**). The composition of elements affects the storage properties of the liquid fuel. Compared to conventional fossil fuels, biomass has a high amount of ~40–50% oxygen content. High oxygen content makes the produced liquid fuel unstable, corrosive and consequently not qualified for transportation and storage [16].

The minor elements present in the biomass material are minerals such as potassium, chlorine, sulfur, silicon, calcium and magnesium [45]. Minerals are present in all biomass species, in a much lower amount than carbon, hydrogen and oxygen elements. Agricultural biomass has much more mineral contents than woody biomass. In a thermal process, minerals turn to ash (**Table 1**). The pyrolysis biochar typically contains up to 90% of the biomass minerals [46]. Ash shifts the size distribution of the char to smaller sizes that make their recovery from the gas stream challenging. An incomplete separation of char and volatiles causes continuous secondary reactions in the liquid phase [16, 47, 48] that accelerates the aging phenomenon and contribute to its instability [49, 50]. Aging phenomena is defined as a slow increase in viscosity bio-oil resulting from secondary reactions [16]. Minerals have a catalytic effect on the rate of secondary reactions [51].

2.2 Pyrolysis products

Products of the pyrolytic reaction are biochar and hot volatile that need to be separated and cooled down quickly. The residence time of char and volatile in direct contact should be less than 2–3 seconds [5, 15, 52]. Char is greatly porous and has a catalytic effect on volatile cracking to non-condensable gases, more char and also non-desirable volatiles [1]. The main product of fast pyrolysis is liquid bio-oil that is obtained by condensation of volatiles, yielding up to 75 wt.% based on a dry feed

Biomass type	Pyrolysis temperature (°C)	Bio-oil yield (wt.%)	Refs.
Pine wood	450	55.0	[36]
Waste furniture sawdust	450	65.0	[37]
Wood sawdust	650	74.0	[38]
Corncob	550	56.8	[39]
Municipal, livestock and wood waste	500	39.7	[40]
Pine wood	450	50	[41]
Rice husks	450	60	[42]
Corn cobs and corn stover	650	61.6	[43]
Sugar cane waste	470	56.5	[44]

Table 2.
Bio-oil yield in fast pyrolysis of different biomass feedstocks.

basis [15]. **Table 3** lists a few key properties of bio-oil and the properties of petroleum grade fuel (Fuel No. 6). Bio-oil has a heating value of about half of conventional fossil fuel oil, 17–19 MJ/kg for bio-oil versus 35–40 MJ/kg for petroleum fossil fuel [52].

The final application of produced bio-oil depends on its quality and properties such as solid content, stability during storage and transportation, oxygen and water contents and its pH value [54]. High oxygen and water content make the bio-oil less stable and less competitive with conventional fossil fuels. Low stabilization, including phase separation and polymerization, and acidic and corrosive properties of bio-oil make its storage and transportation difficult [39, 55]. Based on National Renewable Energy Laboratory's (NREL) recommendation, Multi-Year Program Plan (MYPP 2011), the bio-oil must maintain its consistency for at least 6 months of storage. Based on the published studies, a drastic reduction in water content of bio-oil, insoluble solids and oxygen content and a substantial increase in the calorific value of bio-oil are needed for upgrading to fuel grade.

Gas and biochar are generated as by-products of the pyrolysis process. Optimizing their properties helps to find an application for by-products to make the overall process more economically feasible. They have the potential to provide the heat required for the pyrolysis process. The non-condensable gases of the pyrolysis process have a substantial amount of CO_2 and CH_4. Combustion of these permanent gases makes heat for the pyrolysis process. Biochar has various interesting properties. First one is that the biochar also has a high calorific value. The energy density for charcoal is 9–11 GJ/m^3, while energy density for wood and coal are 8–10 and 25–40 GJ/m^3, respectively [1]. Second, biochar is enriched with carbon that is a good soil amendment. In addition, adding char to soil moves the carbon from atmosphere to soil [56]. Third but not last, pyrolysis biochar has promising adsorption properties for heavy metals [57], dyes [58, 59] and aromatics [60] in agricultural and industrial effluents. The biomass properties influence the biochar properties to be used for adsorption purposes. For example, woody biomass contains more lignin content than agricultural biomass and its biochar has higher surface area than other biomass-derived biochars. Then, it is speculated to have better adsorptivity properties [61].

Characteristics	Fast pyrolysis bio-oil		Heavy petroleum fuel
	Wet	Dry	
Water content (wt.%)	15–25		0.1
Insoluble solids (%)	0.5–0.8		0.01
Carbon (%)	39.5	55.8	85.2
Hydrogen (%)	7.5	6.1	11.1
Oxygen (%)	52.6	37.9	1.0
Nitrogen (%)	<0.1		0.3
Sulfur (%)	<0.05		2.3
Ash (%)	0.2–0.3		<0.1
Heating value (MJ/kg)	17		40
Density (g/ml)	1.23		0.94
Viscosity (cp) @ 50°C	10–150		180

Table 3.
Properties of wood-derived bio-oil and heavy petroleum fuel [53].

3. Feedstock pretreatment

Physical properties like particle size, particle shape and density influence the material handling and flowability of particles that are leading properties for an uninterrupted feeding system [62]. Chemical properties like elemental and proximate analyses influence the pyrolysis products' distribution and properties. Although biomass has inherent heterogeneous properties [1], various physical, chemical and thermal pretreatments contribute to achieving homogeneity in properties of biomass. Homogeneous feedstock with uniform physical and chemical properties plays a key role in a pyrolysis process and quality of its products. Although pretreatments create an additional feedstock cost, they facilitate feeding a wide range of biomass species with broad properties and are beneficial due to producing uniform and homogeneous feedstock for power plants.

3.1 Moisture reduction

Biomass particles are dried for a more-efficient thermal combustion in pyrolysis reactors to produce bio-oil [63, 64]. A fresh biomass has a high moisture content of up to 80% [34]. High moisture content reduces the heating value of the fuel and shifts the ignition point to higher temperatures [6], and inhibits the rise of temperature inside the particles and conversion reactor [65, 66]. Reduced particle and reactor temperatures diminish the liquid yield at the expense of a larger fraction of biochar and non-condensable gases [22, 66]. Water has a catalytic effect on volatile cracking. Di Blasi [67] showed that moisture content conducts the pyrolysis reaction to a low activation energy path that promotes the formation of char, non-condensable gases and more water. Biomass should be dried down to less than 10% moisture content to improve the quality of produced fuel [15]. Pyrolysis of dry material produces less water compared to wet biomass. Bio-oil with a lower amount of water has higher heating value, a lower ignition point, higher combustion rate and a lower potential of phase separation [6].

Rezaei [68] stated that size of particles [69], initial moisture content [70], drying temperature [69, 71], relative humidity of drying gas [72, 73] and particle heating rate [74] influence the rate of moisture loss. Rezaei et al. [75] showed that biomass shrinks during the moisture loss that influences the drying rate. Dehydration of fresh wood causes a reduction in the dimension of wood in a direction normal to the microfibril orientation, whereas the longitudinal shrinkage is usually negligible [76]. Mazzanti et al. [77] showed that the longitudinal shrinkage for poplar wood is negligible. Taylor et al. [78] showed that the radial shrinkage of beech wood is about 70 times of longitudinal shrinkage.

3.2 Size reduction

Grinding reduces the dimensions of a particle and increases the particle's specific surface area (ratio of the particle's surface area to its mass). The same relationship holds for a bulk of particles where the surface area of the solids increases in a given volume of bulk particles. Therefore, the smaller particles have a more exposed surface to the raised-temperature environment to boost the rate of heat and mass transfer between the particles and surrounding. Rezaei et al. [79, 80] showed that larger particles (from 1 to 5 mm) have a delay in heat and mass transfer and reduced rate of pyrolysis (**Figure 2**). On the other hand, grinding the biomass to smaller particles requires more energy input. Rezaei et al. [81] reported that grinding wood chips to produce particles with average sizes of 1, 1.8, 3 and 5 mm requires about 124.0, 85.7, 52.6 and 28.2 kJ/kg, respectively. Van Der Stelt et al. [82] measured

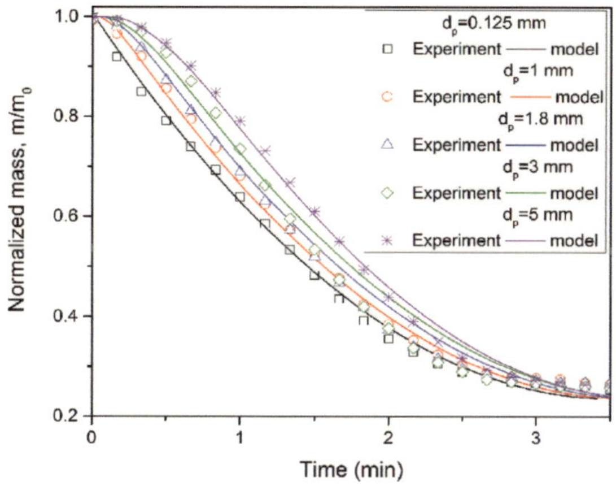

Figure 2.
Effect of particle size on the pyrolysis rate of ground wood particles at 500°C [80].

the power required for size reduction of coal, torrefied woodcutting, willow and demolition wood. He showed that to have particles smaller than 0.6 mm, the grinding power consumption increases sharply. The balance of the conversion rate and energy consumption identifies the optimum particle size appropriate for a fast pyrolysis process.

Figure 3 shows the published data on yields of three phases of solid, liquid and gas in pyrolysis of a range of particle size. In the range of particle size up to 2 mm, only the apricot stone pyrolyzed at 800°C showed a yield that was sensitive to particles size. An increase in particle size increased the char yield at the expense of less liquid yield [15, 22, 26, 83]. Some researchers did not observe any effect of particle size on the yield of products. For example, Şensöz et al. [84, 85] did not observe any meaningful influence of particle size on the pyrolysis products of rapeseed (*Brassica napus* L.) in the range of 0.22–0.85 mm [84] and of debarked pine in the range of 2–5 mm [85]. Encinar et al. [86] reported that liquid yield was independent of biomass particle size during pyrolysis of 0.4–2.0 mm grape residue and olive residue particles pyrolyzed at 500°C. On the other hand, some researchers showed that larger particles in pyrolysis decrease liquid yield and increase biochar yield [22, 83, 87]. NikAzar et al. [22] found out that size increase from 53 to 63 μm to 270–500 μm declined liquid yield from 53 to 38%. He claimed the particle's core temperature diminished by particle size and caused the change in the yield of products. This result was confirmed by some other researchers [15–17, 26, 29, 88].

Most of the published data recommend that particles in the range of 1–2 mm are appropriate for fast pyrolysis [17, 52, 82, 89]. However, the effect of particle size on the pyrolysis process needs more work. The range of particle sizes tested in various studies is seldom comparable and a study on a wider range of particle size seems necessary. In addition, the challenges associated with commercial grinding and feeding the bulky lightweight biomass in the reactor should be taken into account that is out of the scope of the current chapter.

3.3 De-mineralization

The mineral elements in biomass may be listed mostly as potassium, chlorine, sulfur, silicon, calcium and magnesium [45]. The mineral content of biomass exists on the particle surface because of contact with soil during harvest and/or transportation, or within the material as biogenic characteristics.

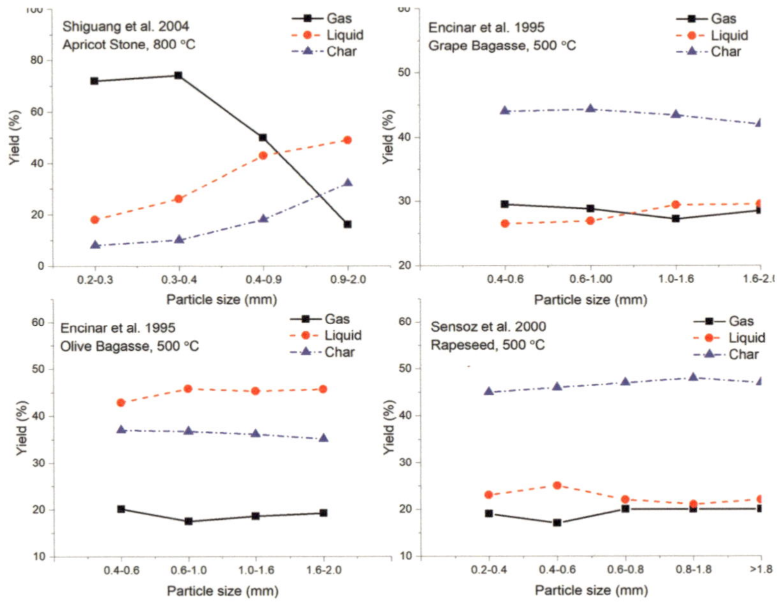

Figure 3.
Effect of biomass particle size on the yield of products in a fast pyrolysis process.

The minerals of the biomass present as ash or fouling in conversion vessels. In a pyrolysis process, the biochar typically contains up to 90% of the biomass minerals [46] that changes the physical properties of char. Ash shifts the size distribution of the char to smaller sizes that hardens its full separation from produced volatile. A partial char separation results in a high solid content bio-oil that cannot be used as turbine fuel [49]. Furthermore, minerals act as a catalyst and cause continuous secondary reactions in liquid phase [16, 47, 48] at the expense of char formation [51, 90]. More secondary reactions promote an increases in bio-oil viscosity with time and accelerate aging phenomenon [50]. Mineral contamination accelerates the catalytic breaking of levoglucosan into unwanted hydroxyl acetaldehyde compounds [51, 91]. Former is desired part of volatiles and latter is an undesirable portion of volatiles. Addition of 0.05 wt.% NaCl to an ash-free cellulose decreases the levoglucosan formation yield by a factor of 6 [92].

One efficient pretreatment to reduce the mineral content is washing the biomass with water, acidic and/or alkaline solutions. Washing with dilute acid and hot water results in a slight decomposition of hemicellulose [93]. Washing with dilute alkali disrupts the lignin structure and solubilize the hemicellulose [94]. Washing biomass prior to pyrolysis takes away a huge amount of minerals from the biomass, up to 70% of the initial minerals [1, 17].

Das et al. [95] studied the effect of various washing solutions, concentrations and the time of washing on ash content and products' yields of sugarcane bagasse pyrolysis (**Figure 4**). The ash content decreased from 1.83% before washing to 0.03%. The only washing solution that had a reverse effect was 5 M HCl. It must have been due to the adding chlorine element into the biomass. The important point is the effect of washing on yield of total liquid versus the bio-oil. The total liquid contains bio-oil and aqueous solutions. All washing solutions reduced the total liquid yield but boosted the bio-oil yield.

Table 4 lists different washing solutions based on demineralization yield, ash content, char and liquid yield and maximum decomposition rate [96]. Washing demineralized the biomass up to 98%, enhanced liquid yield, lowered the char yield, shifted up the decomposition rate and reduced the low molecular weight compounds

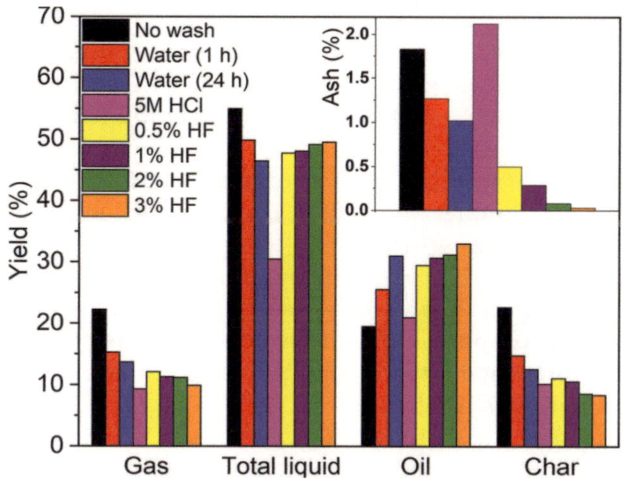

Figure 4.
Effect of de-mineralization pretreatment on biomass ash content and products' yield of sugarcane bagasse pyrolysis [95].

Washing solution	Demineralization yield (%)	Char yield (%)	Volatiles yield (wt.%)	Max. decomposition rate (wt.%/°C)	Tp[3] (°C)
No washing	–	14.9	85.1	0.96	362
HCl[1]	69.3	11.8	88.2	1.55	366
HF[2]	97.3	10.2	89.8	1.15	368
Deionized water	97.7	11.0	89.0	1.23	372
Tap water	98.2	11.4	88.6	1.19	376

[1]*Hydrochloric acid.*
[2]*Hydrofluoric acid.*
[3]*Temperature at which maximum decomposition rate happens.*

Table 4.
Specifications for pyrolysis of demineralized poplar wood at 550°C [96].

in pyrolysis products. Solution temperature also influences the demineralization yield. Deng et al. [7] showed that demineralization of candlenut wood raised from 8% at 30°C to 35% to 90°C. Mineral removal increased the higher heating value from 16.53 to 17.82 MJ/kg and the rate of devolatilization increased too. Natural rain and season of raining change the composition of minerals in biomass [97].

3.4 Torrefaction

Torrefaction is a mild thermal treatment that modifies the structure and chemical composition of biomass by removing hemicelluloses [98], dehydrating and partially reducing cellulose and lignin [99]. Li et al. [13] described torrefaction as a mild heat treatment of biomass at a temperature range of 200–300°C, prolonging 15–30 minutes. Westover et al. [6] divided torrefaction into three stages of non-reactive drying (50–150°C), reactive drying (150–200°C) and destructive stage (200–300°C). During thermal treatment, biomass releases moisture up to the temperature of 150–170°C [11]. In 180–270°C, exothermic hemicellulose degradation happens and biomass turns to brown color. At this stage, torrefaction reaction produces more water, CO_2, acetic acid and phenols [12–14, 35]. The released gases are combustible and may be used to provide a portion of process energy. Beyond 270°C, the reactions are more exothermic and produce CO and some other heavier products such as CH_4 and C_2H_6. Torrefaction continues to a temperature of 300°C, where pyrolysis starts.

Torrefied biomass contains about 80% of the mass and 90% of the energy of the initial biomass [82, 99]. Similar to other thermal processes, mass and energy yield of torrefaction depends on biomass species [99], particle size [14, 26], operating temperature [6, 13, 33, 100, 101] and residence time [13, 33, 101]. Biomass torrefaction has been recognized as a feasible technique to convert raw biomass to a high energy density, hydrophobic, grindable, homogeneous, low moisture content (<5%) [6], better storage stability [34] and low-oxygen-content fuel that is a suitable feedstock for pyrolysis [102].

Torrefaction changes the stiffness and glass transition temperature (T_g) of biomass [6] that contributes to the grindability of the material. Ground torrefied biomass particles are smaller, drier and more uniform in size and physical and chemical properties [13, 14]. **Table 5** lists the specific grinding energy of raw and torrefied pine and logging residue. Higher torrefaction temperature makes the biomass structure more brittle and reduces the specific grinding energy. Specific grinding energy consumption reduced by 90% from 240 kWh/t for raw pine chips to 24 kWh/t for torrefied pine chips at 300°C.

Using the torrefied biomass as a feedstock for fast pyrolysis has various benefits. The level of these modifications depends on torrefaction temperature and residence time. First of all, torrefaction removes the hemicellulose and enriches the biomass into the cellulose and lignin (**Figure 5**). Presence of hemicellulose intensively decreases the yield of levoglucosan and promotes the formation of hydroxyl acetaldehyde. The torrefied feedstock may eliminate or reduce this interaction as it removes hemicelluloses.

Bio-oil of torrefied biomass has a lower water content that increases bio-oil's stability (less phase separation) [100, 104, 105], lower oxygen/carbon ratio and higher calorific value [104] compared to the bio-oil produced from non-torrefied biomass [5, 89, 99, 101]. Because torrefaction extracts contain acidic condensable volatiles such as acetic acid, furfural, formic acid, methanol, lactic acid and phenol from the biomass [106], the bio-oil has a lower acidity and aldehydes content [17, 105].

Sample	Specific grinding energy (kWh/t)	Hemicel. (wt.%)	Sample	Specific grinding energy (kWh/t)	Hemicel. (wt.%)
Untreated-PC[1]	237.7	15.19	Untreated-LR[3]	236.7	14.77
TPC[2]-225°C	102.6	12.87	TLR[4]-225°C	113.8	13.26
TPC-250°C	71.4	6.94	TLR-250°C	110.4	5.87
TPC-275°C	52.0	0.99	TLR-275°C	78.0	5.23
TPC-300°C	23.9	0.56	TLR-300°C	37.6	1.04

[1]*Pine chips.*

[2]*Torrefied pine chips.*

[3]*Logging residues.*

[4]*Torrefied logging residues.*

Table 5.
Specific energy consumption for grinding of untreated and torrefied biomass with a residence time of 30 minutes [98].

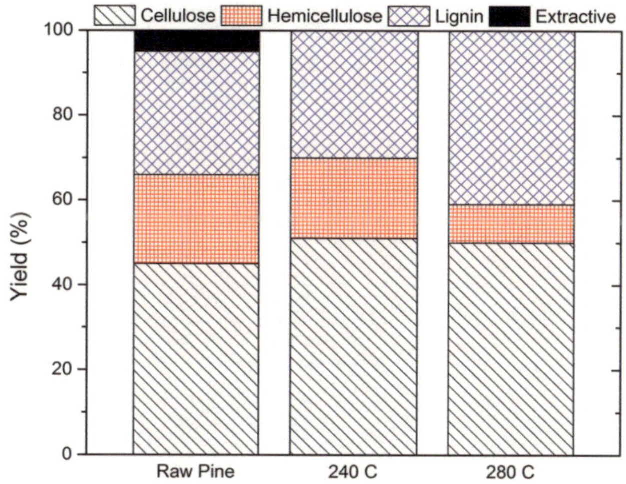

Figure 5.
Composition of raw and torrefied pine at 240 and 280°C [103].

Klinger [105] pyrolyzed the torrefied material and observed 27% water reduction, 36% CO reduction, 55% CO_2 reduction and 67% acetic acid reduction in the produced volatiles compared to volatiles obtained from non-torrefied biomass.

Despite all benefits, pyrolysis of torrefied biomass has a lower yield of total liquid. Liaw et al. [100] reported that pyrolysis total liquid yields of raw Douglas fir and torrefied Douglas fir at 280, 320 and 370°C was 59, 55, 48 and 30%, respectively [100]. Boateng [99] and Zheng [5, 104] reported that more sever torrefaction decreases the total liquid yield and increases the yield of biochar and permanent gases. **Tables 6** and 7 list the results of Zheng's work for pyrolysis of raw and torrefied corncob.

Water content reduced from 35% in bio-oil produced from raw corncob to 21% in bio-oil produced from torrefied corncob at 300°C. The pH increased in the bio-oil prepared from torrefied corncob. The viscosity of bio-oil also increased probably due to a reduction in water content. The acetic acid content decreased moderately with increasing torrefaction temperature and residence time. The furfural content also decreased gradually with torrefaction temperature and residence time. Acetic acid and furfural are mainly derived from hemicellulose and cellulose. Ren et al. [101, 107] conducted a research on Douglas fir in two sequent

Properties and compounds	Raw corncob	Torrefied (240°C)	Torrefied (300°C)
Water content (wt.%)	35.00	30.00	21.00
High heating value (MJ/kg)	14.85	16.49	17.21
pH	2.68	3.30	3.34
Kinematic viscosity @ 20°C (cSt)	3.42	7.27	12.62
Acids (95% acetic acid) (%wb)	7.16	5.65	4.75
Ketones (wt.%)	8.16	7.42	6.02
Furans (wt.%)	0.98	0.8	0.76
Phenols (wt.%)	2.45	4.51	5.27

Table 6.
Physical properties and a few main chemical contents of bio-oil produced from pyrolysis (at 500°C) of raw and torrefied corncob at 240 and 300°C for 20 minutes [104].

Properties	Raw corncob	Residence time: 20 minutes			Temperature: 275°C	
		250°C	275°C	300°C	10 minutes	60 minutes
Ultimate analysis of torrefied corncob (wt.%)						
C	43.87	45.10	48.54	57.86	47.64	57.33
H	6.06	6.20	6.65	6.75	6.10	5.93
O	49.50	48.01	44.06	34.43	45.61	35.92
N	0.53	0.61	0.70	0.86	0.60	0.75
S	0.04	0.08	0.05	0.11	0.04	0.06
O/C	1.13	1.06	0.91	0.60	0.96	0.63
Yield of pyrolysis products (wt.%)						
Bio-oil	57.20	55.15	47.60	40.74	50.36	39.35
Biochar	21.14	24.57	30.21	38.19	26.31	40.70
Non-condensable gas	21.66	20.28	22.19	21.07	23.33	19.95
Physical properties of bio-oil						
Water content	35.0	33.0	25.0	21.0	30.0	22.0
Higher heating value (MJ/kg)	14.85	15.12	16.49	17.21	15.58	17.09
pH	2.68	2.97	3.30	3.34	2.88	3.35
Kinematic viscosity @ 20°C (cSt)	3.42	3.96	7.27	12.62	4.23	12.48
Crystallinity (%)	20.06	25.74	34.82	27.35	23.26	22.83

Table 7.
Effect of temperature and residence time of corncob torrefaction on ultimate analysis of feedstock, the yield of pyrolysis products (at 500°C) and physical properties of bio-oil [5].

articles. They determined the effects of pyrolysis temperature and torrefaction time, as a pretreatment, on the characterization of produced total bio-oil and syngas. Total bio-oil is the sum of condensed liquids from torrefaction and pyrolysis. **Figure 6** shows that bio-oil yield from raw (untreated) biomass ranged from 31 [107] to 53% [101] for pyrolysis at 400 and 450°C, respectively. The bio-oil yield from a 450°C pyrolysis decreased to 51 and 46% for 8 and 15 minutes torrefaction pretreatment at 275°C was carried out. Longer torrefaction reduced slightly the yield of bio-oil. Torrefaction also altered the compositions of syngas by reducing CO_2 and increasing H_2 and CH_4. The syngas produced from pyrolysis step was rich in H_2, CH_4 and CO implying that the syngas quality was significantly improved by torrefaction process [105]. The quantity of syngas increased with the severity of the torrefaction.

Table 8 summarizes a qualitative analysis of benefits of feedstock thermal pretreatment on logistical and quality of bio-oil properties. Column 1 lists major properties that a feedstock would gain as a result of thermal pretreatment. The degree of change in properties depends upon the severity of thermal treatment. Column 2 lists the benefits of these properties on improving conversion efficiency or the quality of produced bio-oil when the pretreated feedstock is pyrolyzed. Column 3 outlines the benefits of feedstock properties from a logistical perspective, that is, handling, transport, storage, unit operations like blending and feeding the feedstock to the pyrolysis reactor. Finally, column 4 lists the potential monetary benefits of thermal pretreatments mostly due to improvements in logistics.

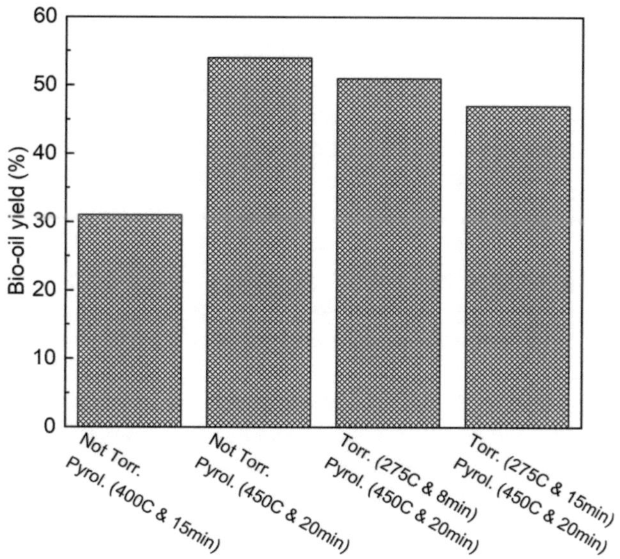

Figure 6.
Bio-oil yield from raw and torrefied Douglas fir at the various duration and temperature [101, 107].

Feedstock properties	Conversion and quality of bio-oil benefit	Logistics benefit	Economic benefit
Grindability	Uniform small size particles for high heat transfer in the reactor	Torrefied particles are dry and a lower tendency for electrostatic charged	Lower investment and operating cost in grinder and grinder operation
Hydrophobicity	Less water content due to a decrease in OH and COOH groups	Store unprotected, exposure to rain, long shelf life	Lower storage cost, lower transport cost
Low moisture content	Reduced water in the bio-oil	Reduced mass to handle, stable in storage	Expensive drying is not required
High heat value	Bio-oil will have a higher heat value due to increased C and low O/C ratio	Can be blended with coal and other high heat value products	Lower cost ($/GJ) No need to design a new combustion chamber
Homogeneity	Predictable conversion performance in the pyrolysis reactor	Reduced quality control, may easily become a commodity	Lower management cost
High density (after grinding)	Can be controlled to a precise particle size and density	Improved flowability, low off-gas emissions	Reduced cost of shipping and storage
Thermal degradation	Lower acidity	Higher energy density (GJ/m^3)	Higher $/GJ
Wet fractionation (wet torrefaction)	Less chlorine and ash	Access to low-quality feedstock (e.g. bark)	Reduced overall feedstock cost ($/t)

These improvements are in comparison with the untreated feedstock.

Table 8.
Overview of benefits of torrefied feedstock for bio-oil production.

4. Conclusions

The energy produced from lignocellulosic biomass is renewable, environmentally friendly and carbon-neutral. Characteristics of bio-oil produced in fast pyrolysis showed that it is capable to substitute conventional petroleum-based fuels. This chapter explained the fast pyrolysis process and the pretreatments that contribute to acquiring a transportation bio-oil with enhanced physical and chemical properties. The modification of biomass feedstock properties through available pretreatment techniques plays a key role in the pyrolysis process. Pretreatments change the biomass feedstock's size, shape, mineral content, composition, hygroscopic properties, homogeneity, grindability, stability and transportability.

Torrefaction as a mild pyrolysis of biomass feedstock is a promising pretreatment that modifies the structure and chemical composition of biomass by removing hemicelluloses, dehydrating and enriching biomass in cellulose and lignin. Torrefaction makes the structure of biomass more brittle that make it more grindable. Grinding torrefied biomass produces more uniform particles with smaller and narrower particle size distribution. Pyrolysis of torrefied material enhances the properties of produced bio-oil. This enhancement is consistent with the DOE MYPP's target that "additional processing of the bio-oil is required to enable it to become a feedstock suitable for use in a petroleum refinery at several entry points." However, this is penalized with a liquid yield reduction.

With all the opportunities of renewable biomass energy, some important issues are still needed to be solved to commercialize the process. To find the least-cost combination of torrefaction along with biomass fast pyrolysis, further testing of the two along with economic analysis are required.

Author details

Hamid Rezaei[1*], Fahimeh Yazdanpanah[1], Jim Choon Lim[1], Anthony Lau[1] and Shahab Sokhansanj[1,2]

1 Chemical and Biological Engineering Department, University of British Columbia, Vancouver, BC, Canada

2 Agricultural and Bioresource Engineering, University of Saskatchewan, Saskatoon, Canada

*Address all correspondence to: hamidrezaei@chbe.ubc.ca

References

[1] Kasparbauer RD. The effects of biomass pretreatments on the products of fast pyrolysis [master of science dissertation]. Mechanical Engineering, Iowa State University; 2009

[2] Carlson TR, Tompsett GA, Conner WC, Huber GW. Aromatic production from catalytic fast pyrolysis of biomass-derived feedstocks. Topics in Catalysis. 2009;**52**:241-252. DOI: 10.1007/s11244-008-9160-6

[3] Biagini E, Narducci P, Tognotti L. Size and structural characterization of lignin-cellulosic fuels after the rapid devolatilization. Fuel. 2008;**87**:177-186. DOI: 10.1016/j.fuel.2007.04.010

[4] Lu H, Ip E, Scott J, Foster P, Vickers M, Baxter LL. Effects of particle shape and size on devolatilization of biomass particle. Fuel. 2010;**89**:1156-1168. DOI: 10.1016/j.fuel.2008.10.023

[5] Zheng A, Zhao Z, Chang S, Huang Z, Wang X, He F, et al. Effect of torrefaction on structure and fast pyrolysis behavior of corncobs. Bioresource Technology. 2013;**128**:370-377. DOI: 10.1016/j.biortech.2012.10.067

[6] Westover TL, Phanphanich M, Clark ML, Rowe SR, Egan SE, Zacher AH, et al. Impact of thermal pretreatment on the fast pyrolysis conversion of southern pine. Biofuels. 2013;**4**:45-61

[7] Deng L, Zhang T, Che D. Effect of water washing on fuel properties, pyrolysis and combustion characteristics, and ash fusibility of biomass. Fuel Processing Technology. 2013;**106**:712-720. DOI: 10.1016/j.fuproc.2012.10.006

[8] Wang Y, He T, Liu K, Wu J, Fang Y. From biomass to advanced bio-fuel by catalytic pyrolysis/hydro-processing: Hydrodeoxygenation of bio-oil derived from biomass catalytic pyrolysis. Bioresource Technology. 2012;**108**:280-284. DOI: 10.1016/j.biortech.2011.12.132

[9] Zanzi R, Sjöström K, Björnbom E. Rapid pyrolysis of straw at high temperature. In: Bridgwater AV, Boocock DGB, editors. Developments in Thermochemical Biomass Conversion. The Netherlands: Springer; 1997. pp. 61-66. DOI: 10.1007/978-94-009-1559-6_4; 10.1007/978-94-009-1559-6_4

[10] Bridgwater AV, Meier D, Radlein D. An overview of fast pyrolysis of biomass. Organic Geochemistry. 1999;**30**:1479-1493. DOI: 10.1016/S0146-6380(99)00120-5

[11] Vlaev LT, Markovska IG, Lyubchev LA. Non-isothermal kinetics of pyrolysis of rice husk. Thermochimica Acta. 2003;**406**:1-7. DOI: 10.1016/S0040-6031(03)00222-3

[12] Uslu A, Faaij APC, Bergman PCA. Pre-treatment technologies, and their effect on international bioenergy supply chain logistics. Techno-economic evaluation of torrefaction, fast pyrolysis and pelletisation. Energy. 2008;**33**:1206-1223. DOI: 10.1016/j.energy.2008.03.007

[13] Li H, Liu X, Legros R, Bi XT, Lim CJ, Sokhansanj S. Torrefaction of sawdust in a fluidized bed reactor. Bioresource Technology. 2012;**103**:453-458. DOI: 10.1016/j.biortech.2011.10.009

[14] Peng JH, Bi HT, Sokhansanj S, Lim JC. A study of particle size effect on biomass torrefaction and densification. Energy and Fuels. 2012;**26**:3826-3839. DOI: 10.1021/ef3004027

[15] Bridgwater AV, Peacocke GVC. Fast pyrolysis processes for biomass. Renewable and Sustainable Energy Reviews. 2000;**4**:1-73. DOI: 10.1016/S1364-0321(99)00007-6

[16] Bridgwater AV. Review of fast pyrolysis of biomass and product upgrading. Biomass and Bioenergy. 2012;**38**:68-94. DOI: 10.1016/j.biombioe.2011.01.048

[17] Isahak WNRW, Hisham MWM, Yarmo MA, Yun Hin T-Y. A review on bio-oil production from biomass by using pyrolysis method. Renewable and Sustainable Energy Reviews. 2012;**16**:5910-5923. DOI: 10.1016/j.rser.2012.05.039

[18] Angın D. Effect of pyrolysis temperature and heating rate on biochar obtained from pyrolysis of safflower seed press cake. Bioresource Technology. 2013;**128**:593-597. DOI: 10.1016/j.biortech.2012.10.150

[19] Altun NE, Hicyilmaz C, Kök MV. Effect of particle size and heating rate on the pyrolysis of silopi asphaltite. Journal of Analytical and Applied Pyrolysis. 2003;**67**:369-379. DOI: 10.1016/S0165-2370(02)00075-X

[20] Antal MJ, Allen SG, Dai X, Shimizu B, Tam MS, Grønli M. Attainment of the theoretical yield of carbon from biomass. Industrial and Engineering Chemistry Research. 2000;**39**:4024-4031. DOI: 10.1021/ie000511u

[21] Jarvis JM, McKenna AM, Hilten RN, Das KC, Rodgers RP, Marshall AG. Characterization of pine pellet and peanut hull pyrolysis bio-oils by negative-ion electrospray ionization fourier transform ion cyclotron resonance mass spectrometry. Energy and Fuels. 2012;**26**:3810-3815. DOI: 10.1021/ef300385f

[22] NikAzar M, Hajaligol MR, Sohrabi M, Dabir B. Effects of heating rate and particle size on the products yields from rapid pyrolysis of beech-wood. Fuel Science and Technology International. 1996;**14**:479-502

[23] Lee J. Biological conversion of lignocellulosic biomass to ethanol. Journal of Biotechnology. 1997;**56**:1-24. DOI: 10.1016/S0168-1656(97)00073-4

[24] Jenkins BM, Kayhanian M, Baxter LL, Salour D. Combustion of residual biosolids from a high solids anaerobic digestion/aerobic composting process. Biomass and Bioenergy. 1997;**12**:367-381. DOI: 10.1016/S0961-9534(97)00086-X

[25] Radlein DSTAG, Piskorz J, Scott DS. Lignin derived oils from the fast pyrolysis of poplar wood. Journal of Analytical and Applied Pyrolysis. 1987;**12**:51-59. DOI: 10.1016/0165-2370(87)80014-1

[26] Basu P, Rao S, Dhungana A. An investigation into the effect of biomass particle size on its torrefaction. The Canadian Journal of Chemical Engineering. 2013;**91**:466-474. DOI: 10.1002/cjce.21710

[27] Thunman H, Leckner B, Niklasson F, Johnsson F. Combustion of wood particles—A particle model for eulerian calculations. Combustion and Flame. 2002;**129**:30-46. DOI: 10.1016/S0010-2180(01)00371-6

[28] Scott DS, Piskorz J, Radlein D. Liquid products from the continuous flash pyrolysis of biomass. Industrial and Engineering Chemistry Process Design and Development. 1985;**24**:581-588. DOI: 10.1021/i200030a011

[29] Demirbas A. Effects of temperature and particle size on bio-char yield from pyrolysis of agricultural residues. Journal of Analytical and Applied Pyrolysis. 2004;**72**:243-248. DOI: 10.1016/j.jaap.2004.07.003

[30] Mani T, Murugan P, Abedi J, Mahinpey N. Pyrolysis of wheat straw in a thermogravimetric analyzer: Effect of particle size and heating rate on devolatilization and estimation of global kinetics. Chemical Engineering Research and Design: Transactions of

the Institution of Chemical Engineers Part A. 2010;**88**:952-958. DOI: 10.1016/j.cherd.2010.02.008

[31] Tsai WT, Lee MK, Chang YM. Fast pyrolysis of rice husk: Product yields and compositions. Bioresource Technology. 2007;**98**:22-28. DOI: 10.1016/j.biortech.2005.12.005

[32] Nassar MM. Thermal analysis kinetics of bagasse and rice straw. Energy Sources. 1999;**20**:831-837. DOI: 10.1016/S0140-6701(99)98986-5

[33] Wannapeera J, Fungtammasan B, Worasuwannarak N. Effects of temperature and holding time during torrefaction on the pyrolysis behaviors of woody biomass. Journal of Analytical and Applied Pyrolysis. 2011;**92**:99-105. DOI: 10.1016/j.jaap.2011.04.010

[34] Tooyserkani Z, Sokhansanj S, Bi X, Lim J, Lau A, Saddler J, et al. Steam treatment of four softwood species and bark to produce torrefied wood. Applied Energy. 2013;**103**:514-521. DOI: 10.1016/j.apenergy.2012.10.016

[35] Grønli MG, Várhegyi G, Di Blasi C. Thermogravimetric analysis and devolatilization kinetics of wood. Industrial and Engineering Chemistry Research. 2002;**41**:4201-4208. DOI: 10.1021/ie0201157

[36] Ortega JV, Renehan AM, Liberatore MW, Herring AM. Physical and chemical characteristics of aging pyrolysis oils produced from hardwood and softwood feedstocks. Journal of Analytical and Applied Pyrolysis. 2011;**91**:190-198. DOI: 10.1016/j.jaap.2011.02.007

[37] Heo HS, Park HJ, Park Y-K, Ryu C, Suh DJ, Suh Y-W, et al. Bio-oil production from fast pyrolysis of waste furniture sawdust in a fluidized bed. Bioresource Technology. 2010;**101**:S91-S96. DOI: 10.1016/j.biortech.2009.06.003

[38] Lédé J, Broust F, Ndiaye F-T, Ferrer M. Properties of bio-oils produced by biomass fast pyrolysis in a cyclone reactor. Fuel. 2007;**86**:1800-1810. DOI: 10.1016/j.fuel.2006.12.024

[39] Zhang H, Xiao R, Huang H, Xiao G. Comparison of non-catalytic and catalytic fast pyrolysis of corncob in a fluidized bed reactor. Bioresource Technology. 2009;**100**:1428-1434. DOI: 10.1016/j.biortech.2008.08.031

[40] Cao J-P, Xiao X-B, Zhang S-Y, Zhao X-Y, Sato K, Ogawa Y, et al. Preparation and characterization of bio-oils from internally circulating fluidized-bed pyrolyses of municipal, livestock, and wood waste. Bioresource Technology. 2011;**102**:2009-2015. DOI: 10.1016/j.biortech.2010.09.057

[41] Thangalazhy-Gopakumar S, Adhikari S, Ravindran H, Gupta RB, Fasina O, Tu M, et al. Physiochemical properties of bio-oil produced at various temperatures from pine wood using an Auger reactor. Bioresource Technology. 2010;**101**:8389-8395. DOI: 10.1016/j.biortech.2010.05.040

[42] Heo HS, Park HJ, Dong J-I, Park SH, Kim S, Suh DJ, et al. Fast pyrolysis of rice husk under different reaction conditions. Journal of Industrial and Engineering Chemistry. 2010;**16**:27-31. DOI: 10.1016/j.jiec.2010.01.026

[43] Mullen CA, Boateng AA, Goldberg NM, Lima IM, Laird DA, Hicks KB. Bio-oil and bio-char production from corn cobs and stover by fast pyrolysis. Biomass and Bioenergy. 2010;**34**:67-74. DOI: 10.1016/j.biombioe.2009.09.012

[44] Islam MR, Parveen M, Haniu H. Properties of sugarcane waste-derived bio-oils obtained by fixed-bed fire-tube heating pyrolysis. Bioresource

Technology. 2010;**101**:4162-4168. DOI: 10.1016/j.biortech.2009.12.137

[45] Jensen PA, Frandsen FJ, Hansen J, Dam-Johansen K, Henriksen N, Hörlyck S. SEM investigation of superheater deposits from biomass-fired boilers. Energy and Fuels. 2004;**18**:378-384. DOI: 10.1021/ef030097l

[46] Raveendran K, Ganesh A, Khilar KC. Influence of mineral matter on biomass pyrolysis characteristics. Fuel. 1995;**74**:1812-1822. DOI: 10.1016/0016-2361(95)80013-8

[47] Chiaramonti D, Oasmaa A, Solantausta Y. Power generation using fast pyrolysis liquids from biomass. Renewable and Sustainable Energy Reviews. 2007;**11**:1056-1086. DOI: 10.1016/j.rser.2005.07.008

[48] Jensen PA, Sander B, Dam-Johansen K. Pretreatment of straw for power production by pyrolysis and char wash. Biomass and Bioenergy. 2001;**20**:431-446. DOI: 10.1016/S0961-9534(01)00005-8

[49] Oasmaa A, Peacocke C, Gust S, Meier D, McLellan R. Norms and standards for pyrolysis liquids. End-user requirements and specifications. Energy and Fuels. 2005;**19**:2155-2163. DOI: 10.1021/ef040094o

[50] Oasmaa A, Sipilä K, Solantausta Y, Kuoppala E. Quality improvement of pyrolysis liquid: Effect of light volatiles on the stability of pyrolysis liquids. Energy and Fuels. 2005;**19**:2556-2561. DOI: 10.1021/ef0400924

[51] Agblevor FA, Besler S. Inorganic compounds in biomass feedstocks. 1. Effect on the quality of fast pyrolysis oils. Energy and Fuels. 1996;**10**:293-298. DOI: 10.1021/ef950202u

[52] Meier D, van de Beld B, Bridgwater AV, Elliott DC, Oasmaa A, Preto F. State-of-the-art of fast pyrolysis in Iea bioenergy member countries. Renewable and Sustainable Energy Reviews. 2013;**20**:619-641. DOI: 10.1016/j.rser.2012.11.061

[53] Doug E. Advancement of bio oil utilization for refinery feedstock. In: Washington Bioenergy Research Symposium. 2010

[54] Sipilä K, Kuoppala E, Fagernäs L, Oasmaa A. Characterization of biomass-based flash pyrolysis oils. Biomass and Bioenergy. 1998;**14**:103-113. DOI: 10.1016/S0961-9534(97)10024-1

[55] Zhang H, Xiao R, Wang D, He G, Shao S, Zhang J, et al. Biomass fast pyrolysis in a fluidized bed reactor under N_2, CO_2, CO, Ch_4 and H_2 atmospheres. Bioresource Technology. 2011;**102**:4258-4264. DOI: 10.1016/j.biortech.2010.12.075

[56] Brewer CE, Schmidt-Rohr K, Satrio JA, Brown RC. Characterization of biochar from fast pyrolysis and gasification systems. Environmental Progress and Sustainable Energy. 2009;**28**:386-396. DOI: 10.1002/ep.10378

[57] Fouladi Tajar A, Kaghazchi T, Soleimani M. Adsorption of cadmium from aqueous solutions on sulfurized activated carbon prepared from nut shells. Journal of Hazardous Materials. 2009;**165**:1159-1164. DOI: 10.1016/j.jhazmat.2008.10.131

[58] Demirbas A. Agricultural based activated carbons for the removal of dyes from aqueous solutions: A review. Journal of Hazardous Materials. 2009;**167**:1-9. DOI: 10.1016/j.jhazmat.2008.12.114

[59] Yang Y, Lin X, Wei B, Zhao Y, Wang J. Evaluation of adsorption potential of bamboo biochar for metal-complex dye: Equilibrium, kinetics and artificial neural network modeling. International

Journal of Environmental Science and Technology. 2014;**11**:1093-1100. DOI: 10.1007/s13762-013-0306-0

[60] Park J, Hung I, Gan Z, Rojas OJ, Lim KH, Park S. Activated carbon from biochar: Influence of its physicochemical properties on the sorption characteristics of phenanthrene. Bioresource Technology. 2013;**149**:383-389. DOI: 10.1016/j.biortech.2013.09.085

[61] Fang Q, Chen B, Lin Y, Guan Y. Aromatic and hydrophobic surfaces of wood-derived biochar enhance perchlorate adsorption via hydrogen bonding to oxygen-containing organic groups. Environmental Science and Technology. 2013;**48**:279-288. DOI: 10.1021/es403711y

[62] Tannous K, Lam PS, Sokhansanj S, Grace JR. Physical properties for flow characterization of ground biomass from Douglas Fir wood. Particulate Science and Technology. 2012;**31**:291-300. DOI: 10.1080/02726351.2012.732676

[63] Vamvuka D. Bio-oil, solid and gaseous biofuels from biomass pyrolysis processes—An Overview. International Journal of Energy Research. 2011;**35**:835-862. DOI: 10.1002/er.1804

[64] Pang S, Mujumdar AS. Drying of woody biomass for bioenergy: Drying technologies and optimization for an integrated bioenergy plant. Drying Technology. 2010;**28**:690-701. DOI: 10.1080/07373931003799236

[65] Ball RC, McIntosh A, Brindley J. The role of char-forming processes in the thermal decomposition of cellulose. Physical Chemistry Chemical Physics. 1999;**1**:5035-5043

[66] Di Blasi C. Modeling intra- and extra-particle processes of wood fast pyrolysis. AIChE Journal. 2002;**48**:2386-2397. DOI: 10.1002/aic.690481028

[67] Di Blasi C. Modeling chemical and physical processes of wood and biomass pyrolysis. Progress in Energy and Combustion Science. 2008;**34**:47-90. DOI: 10.1016/j.pecs.2006.12.001

[68] Rezaei H. Physical and thermal characterization of ground wood chip and ground wood pellet particles [PhD dissertation]. Chemical and Biological Engineering, University of British Columbia (UBC); 2017

[69] Gómez-de la Cruz FJ, Cruz-Peragón F, Casanova-Peláez PJ, Palomar-Carnicero JM. A vital stage in the large-scale production of biofuels from spent coffee grounds: The drying kinetics. Fuel Processing Technology. 2015;**130**:188-196. DOI: 10.1016/j.fuproc.2014.10.012

[70] Berberović A, Milota MR. Impact of wood variability on the drying rate at different moisture content levels. Forest Products Journal. 2011;**61**:435-442

[71] Wan Nadhari WNA, Hashim R, Danish M, Sulaiman O, Hiziroglu S. A model of drying kinetics of acacia mangium wood at different temperatures. Drying Technology. 2014;**32**:361-370. DOI: 10.1080/07373937.2013.829855

[72] Sigge GO, Hansmann CF, Joubert E. Effect of temperature and relative humidity on the drying rates and drying times of green bell peppers (*Capsicum Annuuml*). Drying Technology. 1998;**16**:1703-1714. DOI: 10.1080/07373939808917487

[73] Tapia-Blácido D, Sobral PJ, Menegalli FC. Effects of drying temperature and relative humidity on the mechanical properties of amaranth flour films plasticized with glycerol. Brazilian Journal of Chemical Engineering. 2005;**22**:249-256

[74] Chen D, Zhang Y, Zhu X. Drying kinetics of rice straw under isothermal and nonisothermal conditions: A comparative study by thermogravimetric analysis. Energy and Fuels. 2012;**26**:4189-4194. DOI: 10.1021/ef300424n

[75] Rezaei H, Sokhansanj S, Lim CJ, Lau A, Bi X. Effects of the mass and volume shrinkage of ground chip and pellet particles on drying rates. Particuology. 2018;**38**:1-9. DOI: 10.1016/j.partic.2017.09.001

[76] Mujumdar AS. Handbook of Industrial Drying. Boca Raton: Tylor & Francis; 2006

[77] Mazzanti P, Togni M, Uzielli L. Drying shrinkage and mechanical properties of poplar wood (*Populus alba* L.) across the grain. Journal of Cultural Heritage. 2012;**13**:S85-S89. DOI: 10.1016/j.culher.2012.03.015

[78] Taylor A, Plank B, Standfest G, Petutschnigg A. Beech wood shrinkage observed at the microscale by a time series of X-Ray computed tomographs (Mxct). De Gruyter. 2013;**67**:201-205

[79] Rezaei H, Yazdanpanah F, Lim CJ, Lau A, Sokhansanj S. Pyrolysis of ground pine chip and ground pellet particles. The Canadian Journal of Chemical Engineering. 2016;**94**:1863-1871. DOI: 10.1002/cjce.22574

[80] Rezaei H, Sokhansanj S, Bi X, Lim CJ, Lau A. A numerical and experimental study on fast pyrolysis of single woody biomass particles. Applied Energy. 2017;**198**:320-331. DOI: 10.1016/j.apenergy.2016.11.032

[81] Rezaei H, Lim CJ, Lau A, Sokhansanj S. Size, shape and flow characterization of ground wood chip and ground wood pellet particles. Powder Technology. 2016;**301**:737-746. DOI: 10.1016/j.powtec.2016.07.016

[82] van der Stelt MJC, Gerhauser H, Kiel JHA, Ptasinski KJ. Biomass upgrading by torrefaction for the production of biofuels: A review. Biomass and Bioenergy. 2011;**35**:3748-3762. DOI: 10.1016/j.biombioe.2011.06.023

[83] Li S, Xu S, Liu S, Yang C, Lu Q. Fast pyrolysis of biomass in free-fall reactor for hydrogen-rich gas. Fuel Processing Technology. 2004;**85**:1201-1211. DOI: 10.1016/j.fuproc.2003.11.043

[84] Şensöz S, Angın D, Yorgun S. Influence of particle size on the pyrolysis of rapeseed (*Brassica Napus* L.): Fuel properties of bio-oil. Biomass and Bioenergy. 2000;**19**:271-279. DOI: 10.1016/S0961-9534(00)00041-6

[85] Şensöz S, Can M. Pyrolysis of pine (*Pinus Brutia* Ten.) chips: 1. Effect of pyrolysis temperature and heating rate on the product yields. Energy Sources. 2002;**24**:347-355. DOI: 10.1080/00908310252888727

[86] Encinar JM, Beltrán FJ, Bernalte A, Ramiro A, González JF. Pyrolysis of two agricultural residues: olive and grape bagasse. Influence of particle size and temperature. Biomass and Bioenergy. 1996;**11**:397-409. DOI: 10.1016/S0961-9534(96)00029-3

[87] Scott DS, Piskorz J. The continuous flash pyrolysis of biomass. Canadian Journal of Chemical Engineering. 1984;**62**:404-412

[88] Czernik S, Bridgwater AV. Overview of applications of biomass fast pyrolysis oil. Energy and Fuels. 2004;**18**:590-598. DOI: 10.1021/ef034067u

[89] Meng J, Park J, Tilotta D, Park S. The effect of torrefaction on the chemistry of fast-pyrolysis bio-oil. Bioresource Technology. 2012;**111**:439-446. DOI: 10.1016/j.biortech.2012.01.159

[90] Jendoubi N, Broust F, Commandre JM, Mauviel G, Sardin M, Lédé J.

Inorganics distribution in bio oils and char produced by biomass fast pyrolysis: The key role of aerosols. Journal of Analytical and Applied Pyrolysis. 2011;**92**:59-67. DOI: 10.1016/j.jaap.2011.04.007

[91] Scott DS, Paterson L, Piskorz J, Radlein D. Pretreatment of poplar wood for fast pyrolysis: Rate of cation removal. Journal of Analytical and Applied Pyrolysis. 2001;**57**:169-176. DOI: 10.1016/S0165-2370(00)00108-X

[92] Piskorz J, Radlein DSAG, Scott DS, Czernik S. Pretreatment of wood and cellulose for production of sugars by fast pyrolysis. Journal of Analytical and Applied Pyrolysis. 1989;**16**:127-142. DOI: 10.1016/0165-2370(89)85012-0

[93] Chaiwat W, Hasegawa I, Kori J, Mae K. Examination of degree of cross-linking for cellulose precursors pretreated with acid/hot water at low temperature. Industrial and Engineering Chemistry Research. 2008;**47**:5948-5956. DOI: 10.1021/ie800080u

[94] Alvira P, Tomás-Pejó E, Ballesteros M, Negro MJ. Pretreatment technologies for an efficient bioethanol production process based on enzymatic hydrolysis: A review. Bioresource Technology. 2010;**101**:4851-4861. DOI: 10.1016/j.biortech.2009.11.093

[95] Das P, Ganesh A, Wangikar P. Influence of pretreatment for deashing of sugarcane bagasse on pyrolysis products. Biomass and Bioenergy. 2004;**27**:445-457. DOI: 10.1016/j.biombioe.2004.04.002

[96] Eom I-Y, Kim K-H, Kim J-Y, Lee S-M, Yeo H-M, Choi I-G, et al. Characterization of primary thermal degradation features of lignocellulosic biomass after removal of inorganic metals by diverse solvents. Bioresource Technology. 2011;**102**:3437-3444. DOI: 10.1016/j.biortech.2010.10.056

[97] Bakker RR, Jenkins BM. Feasibility of collecting naturally leached rice straw for thermal conversion. Biomass and Bioenergy. 2003;**25**:597-614. DOI: 10.1016/S0961-9534(03)00053-9

[98] Phanphanich M, Mani S. Impact of torrefaction on the grindability and fuel characteristics of forest biomass. Bioresource Technology. 2011;**102**:1246-1253. DOI: 10.1016/j.biortech.2010.08.028

[99] Boateng AA, Mullen CA. Fast pyrolysis of biomass thermally pretreated by torrefaction. Journal of Analytical and Applied Pyrolysis. 2013;**100**:95-102. DOI: 10.1016/j.jaap.2012.12.002

[100] Liaw S-S, Zhou S, Wu H, Garcia-Perez M. Effect of pretreatment temperature on the yield and properties of bio-oils obtained from the auger pyrolysis of Douglas Fir wood. Fuel. 2013;**103**:672-682. DOI: 10.1016/j.fuel.2012.08.016

[101] Ren S, Lei H, Wang L, Bu Q, Chen S, Wu J, et al. The effects of torrefaction on compositions of bio-oil and syngas from biomass pyrolysis by microwave heating. Bioresource Technology. 2013;**135**:659-664. DOI: 10.1016/j.biortech.2012.06.091

[102] Tumuluru JS, Sokhansanj S, Wright CT, Boardman RD. Biomass Torrefaction Process Review and Moving Bed Torrefaction System Model Development. US: Department of Energy, O.o.B.P., editor. Idaho National Laboratory; 2010

[103] Liu Q, Zhong Z, Wang S, Luo Z. Interactions of biomass components during pyrolysis: A Tg-Ftir study. Journal of Analytical and Applied Pyrolysis. 2011;**90**:213-218. DOI: 10.1016/j.jaap.2010.12.009

[104] Zheng AQ, Zhao ZL, Chang S, Huang Z, He F, Li HB. Effect of

torrefaction temperature on product distribution from two-staged pyrolysis of biomass. Energy and Fuels. 2012;**26**:2968-2974. DOI: 10.1021/ef201872y

[105] Klinger JL. Production of biofuel intermediates from woody feedstocks via fast pyrolysis and torrefaction [master of science dissertation]. Chemical Engineering, Michigan Technology University; 2011

[106] Prins MJ, Ptasinski KJ, Janssen FJJG. More efficient biomass gasification via torrefaction. Energy. 2006;**31**:3458-3470. DOI: 10.1016/j.energy.2006.03.008

[107] Ren S, Lei H, Wang L, Bu Q, Chen S, Wu J, et al. Biofuel production and kinetics analysis for microwave pyrolysis of douglas fir sawdust pellet. Journal of Analytical and Applied Pyrolysis. 2012;**94**:163-169. DOI: 10.1016/j.jaap.2011.12.004

Section 2

Applications of Biomass in Biofuel

Chapter 5

Agro-Industrial Waste Revalorization: The Growing Biorefinery

Flora Beltrán-Ramírez, Domancar Orona-Tamayo, Ivette Cornejo-Corona, José Luz Nicacio González-Cervantes, José de Jesús Esparza-Claudio and Elizabeth Quintana-Rodríguez

Abstract

Agro-industrial residues have been the spotlight of different researches worldwide, due to some of their constituents being raw material to generate a diversified variety of industrial products. Nowadays, this situation keeps prevailing and will increase continuously in the future. In the agroindustry, diverse biomasses are subjected to distinct unit processes for providing value to different waste materials from agriculture, food processing, and alcoholic industries. In this chapter, we reported an updated survey of different renewable organic materials that including agricultural wastes can be converted to bioenergy. Similarly, these wastes encrypt different bioactive compounds with an excellent nutraceutical functions and with high adding value. In addition, biocomposites can be elaborated using fibers from wastes with a wide variety of applications in the automotive and packaging industry. Vinasses derived from tequila industry in Mexico represent a lot of potential to extract biocompounds, and we propose a process to obtain them. A perspective of market trend is mentioned in this chapter for compounds derived from agro-industrial wastes. Adding value to those agro-industrial wastes can provide the reduction of negative impact emission, discharge, or disposal, solves an environmental problem, and generates additional income.

Keywords: bioenergy, biocomposites, bioactive peptides, biocompounds, vinasses

1. Introduction

Agro-industrial residues provide an enormous potential to generate sustainable products and bioenergy. An integrated biorefinery is turning into a promising solution with multiple outputs (biofuels, bioactive biocompounds, and biomaterials). Most of the residues generated are intended for landfill or are disposed in an uncontrolled way, causing environmental damage and economic loss. For that reason, it is necessary to develop a sustainable management of them. An integral waste management is proposed in the concept of circular economy to exploit renewable resources. Circular economy is based on the concept of biorefinery and the approach to reduce, reuse, and recycle waste with the objective to recover materials derived from waste considering them as renewable resources [1].

A wide range of metabolites, materials, and energy can be obtained through the exploitation of agricultural residues. Being the commercial-scale technologies is the bottleneck to produce marketable bioproducts. In this review, we put aboard the multiple outputs that can have the agro-industrial residues from bioenergy until a wide array of metabolites can be extracted. In addition, we investigate the potential market for some products derived from residues, searching the revalorization of them.

2. Production of bioenergy

Nowadays, due to the increase in population, it is necessary to find a sustainable solution for the enhanced demand of energy in the world. The fossil fuels are limited and nonrenewable resources; the use of biomass for energy production seems to be a solution to provide energy and reduce the dioxide carbon emissions. The term biomass includes energy crops, residues, and other biological materials that can be used to produce renewable energy [2]. The first-generation biofuels are produced from agricultural crops such as corn, sugarcane, soybean oil, and sunflower [3]. However, there is a conflict due to these biomasses being used for food and generating the name "food versus fuel" [4]. Additionally, emissions of greenhouse gases (GHG) are believed to be lesser for second-generation biofuels than the first-generation fuels [3]. For these reasons, agro-industrial residues have gained attention due to their disponibility, and they include residues from crop, food, and oil industries.

2.1 Solid fuels

Pellets are the most common solid biofuels used; they are cylindrical structures made by compression derived commonly from agricultural residues, forest products, and wood industries [5]. Pellets are used mainly for house heating and in industrial sector. Even though the agro-industrial residues have less energy content than fossil fuels, their use presents great advantages such as the reduction of logistic costs, easy storage, and provision of a great opportunity for the revalorization of these unused residues [5]. For the pellet elaboration, the biomass is treated to be compacted and densified; this includes the drying, and after, the biomass is milled to obtain particles with similar size [6]. Afterward, the material is pressed in a pelletizer and pellets are packaged and stored. Some common methods to improve energy density are torrefaction, steam explosion, hydrothermal carbonization, and biological treatment. In torrefaction, reactions of dehydration and decarboxylation occur lowering proportions in O/C and C/H and increasing heating value [7]. Steam explosion is a treatment with hot steam under pressure and followed by decompression which disintegrates the lignocellulosic structure [8]. Steam explosion treatment increased the heating value in pellets from a different biomass [9]. The international market of pellets derived from wood has been increased, the USA, Canada, and Russia being the largest exporters to Europe, which is the main consumer in the world [10]. Several applications and uses have the pellets from residential to large-scale power plants. The growing demand of sustainable and renewable fuels places the agro-industrial residue pellets with a great potential to supply renewable energy.

2.2 Liquid fuels

Liquid fuels as diesel and petrol are being replaced by liquid biofuels as biodiesel, bio-oil, bioethanol, and butanol. Biodiesel is obtained from feedstock oil as waste cooking and frying oil, animal fats, and fish and microalgae oil, leather, winery, and agro-industrial wastes, directly or indirectly. Oleaginous microorganisms are used in the indirect way for biodiesel production; lipids produced are extracted to be transformed to biofuel. For biodiesel production, three main steps are included: pretreatment, transesterification, and separation. Pretreatment allows agro-industrial residues to be assimilable for the microorganisms and is categorized into acidic, basic, thermal, enzymatic, or combination treatments [11]. Other important aspect to consider is that during pretreatments inhibitors for microbial growth as furfural, acetate, and others can be formed and are necessary to find tolerant strains or medium detoxification [12].

Bio-oils are obtained from biomass through two main processes: pyrolysis and liquefaction [13]. Pyrolysis has taken more attention; fast pyrolysis of lignocellulose biomass for bio-oil production is low cost compared to liquefaction that produces low yield at high cost [14]. Due to their physicochemical characteristics, bio-oils cannot be used for fuel applications without previous treatment [13]. Treatments are based in partial or total elimination of oxygen, and two catalytic routes have been proposed: cracking and hydrotreating. Pyrolysis of agro-industrial residues has been reported for sesame, mustard, *Jatropha*, palm kernel, cottonseed, and neem oil cakes showing an additional value for these residues and reducing wastes [15].

Bioethanol is the most common biofuel, and their production involves steps as pretreatment, saccharification, fermentation, and distillation [16]. Pretreatment allows cellulose to unwind from hemicellulose and lignin to be more available for enzymatic hydrolysis, and commonly physical, chemical, and biological treatments are used to achieve this purpose [17]. The enzymatic hydrolysis allows converting cellulose to glucose or galactose monomers and presents a low toxicity as well as low utility cost and corrosion compared to chemical hydrolysis [18]. Biological treatment is an alternative to liberate cellulose with the use of microorganisms mainly as brown-rot, white-rot, or soft-rot fungi [19]. Once the saccharification is obtained, fermentation is carried on with microorganisms able to produce ethanol. For the microorganism selection, some parameters are necessary to have a broad-substrate utilization that is derived in a high ethanol yield and productivity, to be tolerant to high ethanol concentrations, temperature, and inhibitors presented in hydrolysate for which genetically modified or engineered microorganisms are a good option to achieve a complete utilization of sugars and better production [17]. The simultaneous saccharification and fermentation (SSF) and the separate hydrolysis and fermentation (SHF) are the most common processes usually used to ethanol production [16]. SSF using olive pulp from oil extraction and the yeast *Kluyveromyces marxianus* showed ethanol yields of 76% [20].

Due to its higher heat of combustion and less volatility and it being mixed with gasolines in higher percentage without any modifications in the car engines, butanol is considered a promising renewable biofuel [21]. Butanol is produced through anaerobic biological fermentation process using the *Clostridia* genus [22]. Agricultural residues can be used for economical production of butanol. Simultaneous hydrolysis of wheat straw to sugars and fermentation to butanol resulted in an attractive option for ABE fermentation [23]. Rice bran has resulted to be an effective substrate to butanol production using *C. saccharoperbutylacetonicum* [24]. Agricultural residues can be a promissory source to be efficiently utilized as substrate for butanol production.

2.3 Gas fuels

Biobutanol is a product from anaerobic biological process called ABE fermentation, which converts sugar by using genus *Clostridia* into butanol, acetone, and ethanol in a ratio of 6:3:1, respectively. In this process, genus *Clostridia* such as *Clostridium acetobutylicum*, *Clostridium beijerinckii*, *Clostridium saccaroperbutylacetonicum*, and *Clostridium saccharoacetobutylicum* showed significant activity for synthesis of butanol with higher yield. Biobutanol is a product from anaerobic biological process called ABE fermentation, which converts sugar by using genus *Clostridia* into butanol, acetone, and ethanol in a ratio of 6:3:1, respectively. In this process, genus *Clostridia* such as *Clostridium acetobutylicum*, *Clostridium beijerinckii*, *Clostridium saccaroperbutylacetonicum*, and *Clostridium saccharoacetobutylicum* showed significant activity for synthesis of butanol with higher yield.

Biobutanol is a product from anaerobic biological process called ABE fermentation, which converts sugar by using genus *Clostridia* into butanol, acetone, and ethanol in a ratio of 6:3:1, respectively. In this process, genus *Clostridia* such as *Clostridium acetobutylicum*, *Clostridium beijerinckii*, *Clostridium saccaroperbutylacetonicum* and *Clostridium saccharoacetobutylicum* showed significant activity for synthesis of butanol with higher yield. Biobutanol is a product from anaerobic biological process called ABE fermentation, which converts sugar by using genus *Clostridia* into butanol, acetone, and ethanol in a ratio of 6:3:1, respectively. In this process, genus *Clostridia* such as *Clostridium acetobutylicum*, *Clostridium beijerinckii*, *Clostridium saccaroperbutylacetonicum*, and *Clostridium saccharoacetobutylicum* showed significant activity for synthesis of butanol with higher yield.

Lignocellulosic biomass is a potential source of glucose, xylose, mannose, and arabinose and other organic compounds that can be anaerobically degraded to produce biogas [25]. Biogas is produced through an anaerobic digestion with four steps identified as hydrolysis, acidification, and production of acetate and finally methane using a microorganism consortium [26]. The final product is a gas mixture composed mainly of methane and carbon dioxide and traces of hydrogen sulfide, ammonia, hydrogen, and carbon monoxide [27]. For the enhancement of biogas production, it is necessary to apply pretreatments, and the most commonly used are dilute acid hydrolysis, steam explosion, alkaline hydrolysis, and liquid hot water [28], while Song et al. tested nine pretreatments showing that H_2O_2 and $Ca(OH)_2$ enhance methane yields [29].

3. Biocompounds from agro-industrial wastes

3.1 Polyphenols

Phenolic compounds are a group of chemical compounds that are widely distributed in nature, and their basic structure varies from a simple molecule to a complex skeleton and hydroxyl substituents. These compounds are being the most desirable phytochemicals due to their antioxidant activities that can be useful for the control of different human diseases or disorders [30]. Due to their reactivity, these compounds efficiently interact with important biomolecules such as DNA, lipids, proteins, and other cellular molecules to produce desired results, which then are used for designing natural therapeutic agents. Flavonoids, tannins, anthocyanins, and alkaloids are polyphenols with industrial significance and are present in fruits and plants. In addition, most of the phenolic complexes are found in barks, shells, husk, leaves, and roots [30]. Recently, agro-industrial wastes from fruits, vegetables, and crops have been subjected to different metabolite methods of extraction as a potential source of industrial bioactive compound production. For example, tomato processing industry approximately generates 8.5 million of tons of wastes globally [31], wastes such as seeds, pruning, and peels which contain a high concentration of bioactive phytochemicals. In that sense, peels and seeds of tomatoes are a richer source of bioactive compounds such as carotenes, terpenes, sterols, tocopherols, and polyphenols [32], which exhibited excellent antimicrobial and antioxidant activities and high support of dietary fiber. Other important crop that generates a high amount of waste is the coffee production. Due to the heterogeneous nature of coffee waste, most of the authors are investigating its possible revalorization to determine the content of chemical compounds such as tannins and phenolic compounds. Exhausted and spent coffee ground wastes derived from industries, restaurants, and domestics are a valuable source of phenolic compounds. For example, in coffee waste derived from coffee industries, different ranges of concentrations of polyphenols and tannins around of 6 and 4%, respectively, were found [33]. *J. curcas* and *Ricinus communis* are the most important energetic plants for the biofuel industries; these plants generated high amounts of residues such as seed cake, pruning material, and seed shells with high concentrations of bioactive compounds. In fact, shells of this plants contained high contents of phenolic compounds and exhibited strong antioxidant activities [34]. Extracts of residual wastes of seeds, leaves, fruits, stems, and roots derived from *R. communis* exert different nutraceutical effects such as antioxidant, antimutagenic, as well as DNA protection against photooxidative stress [35].

3.2 Pigments

Agro-industrial wastes can be used as a feedstock extraction and for different fermentation processes as a main source of microbial nutrients to produce biopigments useful in food and cosmetic industries. Chemically synthesized food colorants used as the additives in foods cause the risk of toxicity and hazardous effects to the consumers, than the natural pigments, that are quite safer, nontoxic, and nonhazardous for the environment [36]. The production of natural pigments can be derived from direct plant extraction (e.g., anthocyanins, chlorophylls, carotenoids, and melanin) or by fermentative production through the cultivation of bacteria, yeast, fungi, and algae (e.g., phycocyanins, xanthophylls, and melanin) [37]. Cyanobacteria and microalgae produce high amounts of beta-carotene and astaxanthin, which are used in the industries and have a great commercial value in pharmaceutical and food industries [38]. Different microorganisms such as *Streptomyces*, *Serratia*, *Cordyceps*, *Monascus*, *and Paecilomyces*, *Penicillium atrovenetum*, *Penicillium herquei*, *Rhodotorula*, *Sarcina*, *Cryptococcus*, *Phaffia rhodozyma*, *Pseudomonas, Bacillus sp.*, *Vibrio*, *Monascus purpureus*, *Achromobacter*, *Yarrowia*, and *Phaffia* have shown their potential in pigment production as a major source of blue and yellow-red pigments [39]. Other important pigment is the melanin, which is present in animals, plants, and microorganisms to provide stress protection against UV radiation, oxidation, and defense [40]. This pigment is used for the cosmetic and pharmaceutic industries with a photoprotective and antioxidant importance in different products. The use of agro-industrial wastes such as fruits is a potential source for the melanin biosynthesis by microorganisms and is an attractive choice for commercial-scale production. For example, fruit, wheat bran extracts, and cabbage wastes were used as inoculum in *Bacillus safensis* [41], fungus *Auricularia auricula* [42], and *Pseudomonas* sp. [43] for the melanin production. Melanin is specially found in the seed coat of different plants; however, it is also found in other plant structures such as black spots of leaves, flowers, and seeds [44]. There are a few reports related to the melanin extraction from agro-industrial wastes. In that sense, sunflower husk derived from the oil production was subjected to the melanin extraction, and a technological scheme of melanin production from this waste was developed with a potential application as prophylactic mean and medicinal agent for the treatment of human diseases [45]. Similarly, residues as shells and epicarp from walnut contain high amounts of melanin with a high antioxidant capacity [46].

3.3 Peptides

Bioactive peptides are encrypted within the protein sequences with different bioactivity functions and relevant in some important disorders in human health such as cancer, hypertension, antioxidant functions, diabetes mellitus, and other important diseases. These peptides may have different sizes, around 2–20 amino acid residues per molecule with molecular masses between 1 and 6 kDa and based on their physical properties and amino acid composition [47] which make them very attractive for different applications in pharmaceutical and food industries. Waste can contain many valuable substances, and through a suitable process or technology, this material can be converted into value-added products or raw materials that can be used in secondary processes. Residual wastes generated by agro-industries are a protein-rich source and have become an alternative for obtaining compounds with bioactivity, mainly from protein hydrolysates; their extraction processes do not involve negative environmental impacts [48]. The principal residual wastes generated by the agro-industrial activities are soybean meal, residues of oiled plants, and rapeseed meal [48]. Those peptides can generate in the market peptides and protein drugs more than $40 billion/year, with an accelerated pace in the drug market [48]. The press cake, after oil extraction from *J. curcas* (not toxic genotypes) in biodiesel production, represents a potential of new source of protein for food and feed uses. The seed cake of *Jatropha* contains a high concentration of storage proteins mainly glutelins and globulin fractions [49] that encrypted peptides with antioxidant, chelating, and antihypertensive activities [50]. Some peptides have activities against bacteria that can reduce the human infections. In that sense, a trypsin inhibitor was purified from castor bean waste of seed cakes; the 75-kDa peptide displayed antibacterial activity against *Bacillus subtilis*, *Klebsiella pneumoniae*, and *Pseudomonas aeruginosa*, which are important human pathogenic bacteria. In addition, microscopy studies indicated that this peptide disrupts the bacterial membrane with loss of the cytoplasm content and ultimately bacterial death. The author concludes that this peptide is a powerful candidate for the development of an alternative drug that may help reduce hospital-acquired infections [51]. Other important seed cakes from oiled plants can be used for the peptide characterization. For example, chia (*Salvia hispanica*) seed cake is novelty for the peptide extraction; the seed cake contains high amounts of proteins that encrypted different peptides with antioxidant, antidiabetic, and antihypertensive activities [46].

4. Biocomposites

Biocomposites are formed by a polymer matrix and natural fibers, which act as reinforcements. There are six types of natural fibers commonly used in biocomposite elaboration: grass and reed fibers (wheat, corn, and rice), core fibers (kenaf, jute, and hemp), bast fibers (jute, flax, hemp, ramie, kenaf, bamboo, and banana), seed fibers (coir, cotton and kapok), leaf fibers (abaca, sisal, and pineapple), and other types (wood and roots) [52].

The composition of natural fibers consists mainly of cellulose, hemicellulose, and lignin. Cellulose in plants is the main component that provides stability and strength to the cell walls, and this component directly influences the biocomposite production for a defined application, whether in the textile, automotive, and others. Lignin is a highly cross-linked structure, and the amount of this directly influences the structure, properties, morphology, hydrolysis rate, as well as the flexibility of the fibers. Besides, fibers with greater amount of lignin have less amount of cellulose, and this will also depend on the application of the fiber.

The fibers can be used in both thermoplastics and thermosets. Thermoplastic matrices include polypropylene (PP), high-density polyethylene (HDPE), polystyrene (PS), and polyvinyl chloride (PVC). Thermosets include epoxy, polyester, and phenolic resins. In recent years, the number of studies focused on these materials has increased, because they are environmentally friendly and have low production costs, easy workability, good properties of lightness, mechanical strength, and thermal insulation [53]. However, due to the hydrophilic nature of natural fibers and hydrophobic nature of polymer matrix, there is no good interfacial interaction between the two materials, and therefore, the mechanical properties are deteriorated. Based on the above, chemical and physical treatments have been developed to modify the surface of natural fibers and promote interfacial adhesion with the polymer matrix [54].

Among the chemical treatments stands out alkali, benzoylation, cyclohexane, silicon, peroxide, acetylation, sulfuric acid, stearic acid treatment, and the modification with maleic anhydride. The chemical modification provides more dimensional stability and reduces water absorption capacity [55]. Alkaline treatment is the most used and consists in eliminating the lignin, wax, and oil of the fibers, since these components act as a barrier between the polymeric matrix and the fibers; and in turn, it is possible to increase the roughness in the surface of the fibers [56]. Another alternative to improve the compatibility between these materials is using compatibilizing agents, such as maleic anhydride grafted with polyolefins, either polypropylene or high-density polyethylene. The main factors that affect the processing and performance of biocomposites are the presence of moisture, type, shape (short or long), concentration, and orientation of the fibers. The processing method for obtaining biocomposites will depend on the type of fiber, for example: twin-screw extruder and hydraulic press, injection molding, melt mixing, and single-screw extruder for short-fiber-reinforced composites [57]. New technologies can improve the processing of these materials to make it easier.

The main applications of the biocomposites are automotive parts, packaging, military industry, aerospace, medical articles, etc. The interest of the automotive sector in developing biocomposites lies mainly in reducing the consumption of fiber glass because it is more expensive than the natural one and, in turn, making the vehicles lighter, and it also contributes to the consumption of less combustible and to the fact of being eco-friendly. In recent years, Toyota, Mercedes-Benz, Ford, Mitsubishi, and Daimler Chrysler AG have incorporated biodegradable materials in the exterior parts of some of their vehicles [58]. Pracella et al. [58] studied the functionalization, compatibilization, and properties of polypropylene (PP) composites with hemp fibers. The fibers were functionalized with glycidyl methacrylate (GMA). PP/hemp composites at various compositions were prepared in a Brabender internal mixer. All modified composites showed improved fiber dispersion in the polyolefin matrix and higher interfacial adhesion with respect to the unmodified PP/hemp. Composites showed an increase in Young modulus as compared to PP due to the addition of PP-g-GMA. Vilaseca et al. [59] studied the effect of alkali treatment on interfacial bonding in abaca fibers. They used an epoxy resin, and the results showed that alkali treatments modify the structure and chemical composition of abaca fibers. Abaca fibers treated in 5 wt. % NaOH showed excellent interfacial adhesion with epoxy resin. Bledzki et al. [60] carried out polypropylene-based biocomposites with different types of natural fibers (jute, kenaf, abaca, and softwood) to compare their performance under the same processing conditions, and they found that the properties of biocomposites depend on geometry of the fibers. Kenaf provides strength to biocomposites, abaca obtains the best results in impact resistance, jute fibers are the most stable thermally, and the wood microfibers have good resistant strength. Currently, several studies have been carried out with other types of fibers, in which we can mention agave, castor plant, and *J. curcas* fibers [61]. During the tequila production process, large amounts of waste are produced (mostly fiber), and in the case of castor plant and *J. curcas*, only the seeds are used for the extraction of oils, and the rest of plant is discarded. Therefore, an alternative to take advantage of this waste is to use it to develop biocomposites. Zuccarello et al. [62] demonstrated that the agave variety plays an important role on the mechanical performance of the fibers and they proposed an innovative and eco-friendly method for the fiber extraction based on the simple mechanical pressing of the leaves, alternated to proper water immersions avoiding alkaline treatment. They used an eco-friendly green epoxy and a polylactic acid (PLA) to obtain renewable biocomposites. In another work, Zuccarello et al. [62] studied the effect of agave fiber size on epoxy resin and PLA composites. This study showed that biocomposites with short fibers fail to act as a reinforcement, while the long fibers in the compounds with PLA achieve a high mechanical strength. Vinayaka et al. [63] elaborated composites with polypropylene and fibers extracted from the outer layer of *R. communis* (castor plant), which exhibited an elongation at 5% that was higher than the common bast fibers jute and flax, and the strength at 350 MPa was similar to that of jute but lower than that of cotton. Biocomposites have an enormous potential of applications and a growth market especially in automotive industry.

5. Vinasse and tequila production in Mexico, a case study

The production of alcohol and alcoholic beverages such as wine, beer, and tequila generates two main residues in its process, one of them is the solid part called bagasse and a liquid part obtained from the distiller which is known as vinasse [64]. In Mexico, more than 70% of the establishments that produce vinasse come from the production of tequila and mezcal, 20% from beer, and the rest from the production of wines from grapes and other fruits. Due to the denomination of origin of tequila and mezcal, researchers consider as essential the study of the vinasse process, due to the high environmental and economic impacts in Mexico. In recent years, to decrease the high volume of residual vinasses, it has been decided to generate compost from bagasse and vinasse [65], which is given to farmers to use it in their crops; however, the production of compost uses less than 50% of the vinasses, and the rest is discarded without treatment. The tequila production generates high volumes of vinasses on a ratio of 10–12 L for each liter of tequila produced; they have a high organic content that causes damage to ecosystems by anoxia and acidification of water. Biodigestion systems have been developed for the removal of solids [66]. Nevertheless, the treatments are expensive and with a low efficiency, which has not been achieved an industrial implementation.

Due to the physicochemical characteristics of vinasses, these represent a high source of contamination that must be contained and treated to avoid serious damage to ecosystems [67]. Due to the current production volumes of tequila, it is reported that during 2017 more than 271 million liters were produced (https://www.crt.org.mx/) of which between 2710 and 3252 million liters of vinasse would be obtained [68]. For elaboration of the tequila, two main industrial processes are used to produce the fermentation juice. In the first one, the agave is cooked and squeezed to obtain the juice, and there is a subsequent fermentation process. The most recent and apparently energy efficient is to squeeze the raw agave heart by spraying it with a little hot water for this juice, to be used later in the fermentation process. Vinasse components have been compared between both processes showing significant changes where cooking processes present the highest contents of organic acids compared with spraying with steam [69].

We carry out the characterization of the compounds present in the vinasse that leave the distillation process in order to identify the majority of compounds and to propose an adequate purification process that allows to preserve the properties, as well as to avoid their degradation. The components of the vinasse vary between the different fermentation processes, since they depend on the raw materials; the cooking time will give the characteristics of the juices or liqueurs, the fermentation process, and the variables of distillation, so it will be necessary to carry out a characterization study for the vinasse. Since the vinasse has been treated as a residue, no measures are taken to prevent its degradation or contamination. It is important to obtain reliable results, collect fresh vinasse and free of foreign contaminants, and keep it cool, clean, and not exposed light to prevent its degradation. Vinasse is a complex mixture in which organic compounds with very different chemical characteristics are found. In addition, oils and fats that are contained solubilize another group of compounds, making this mixture difficult to separate.

It is required to carry out an integral development that allows the correct treatment of each component of the vinasse. Three general stages of separation are proposed using as model the composition of the vinasse previously characterized (**Figure 1**). The first stage is the separation of solids and liquids; the second a separation of water-soluble compounds such as alcohol traces and polar compounds of solids belonging to organic matter generated during fermentation and some other solids from the broth culture as carbohydrates, proteins, and mineral salts, among others; and finally the generation of value-added products, organic compounds, and solids of high organic matter content, which can easily be recycled in the form of fertilizer for industrial use.

Each stage requires the implementation of independent processes, but they will guarantee the obtaining of water that is easy to use in compliance with current regulations. The feasibility of obtaining compounds of antioxidant capacity of commercial interest derived from vinasse represents an income that costs the entire process.

As mentioned previously, currently in Mexico the treatment of vinasse has been limited mainly due to the fact that there is no solvency or economic technology that companies can implement. Different types of treatment are being used that depend mainly on the size of the company and the total volume of production; **Table 1** summarizes the strategies used to treat vinasse.

Pretreatments and primary treatments are the most commonly used, as they are economical and simple to implement on any scale. Pretreatments are useful only for acidity reduction but do not eliminate organic load or color. Sedimentation ponds allow the removal of 80% of the sedimentable solids; however, it does not reduce the organic load or fats. The flotation strategy consists of applying air together with

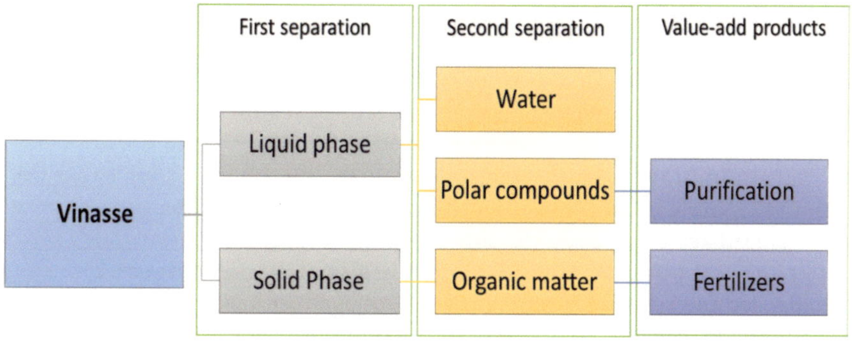

Figure 1.
General process to recovery of compounds from vinasse obtained from distillation processes.

Classification	Treatment	Application	Advantage
Pretreatment	Temperature low	Pool circulation	Economic and popular
	pH neutralization	$Ca(OH)_2$ addition	Economic and popular
Primary treatment	Sedimentation pools	Storage	Easy for industrial volumes
	Air flash floating	Polymer addition	Easy for industrial volumes
Physicochemical treatment	Coagulation	$Al_2(SO_4)_3$ addition	Good remotion of solids
	Flocculation	Cationic polymer addition	Good remotion of solids
Biological	Anaerobic fermentation	Biodigestion	Methane generation
	Acidogenesis	Biodigestion	H_2 and CO_2 generation
New treatments	Oxidation	Redox reaction using ozone, H_2O_2, UV radiation, or Cl	Remotion of color, odor, and organic matter

Table 1.
Treatments actually used for the disposal of vinasse.

a polymer that allows accelerating the separation of soluble solids, and it is used as preparation for a biological process. The process is useful for the removal of solids, it is not efficient in the chemical demand of oxygen (CDO), and in addition, it is expensive in industrial scales. On the other hand, these strategies represent a focus of soil and subsoil contamination by filtration. The physicochemical treatments are mostly used on a pilot scale and generally used in two stages; in one, a coagulant is added to agglomerate soluble solids and then a flocculant for remotion. It is efficient in 20–30%, and its ability is being studied to remove the color; at the laboratory level, it has achieved a 70% reduction in color and 30% in CDO. Although there are already reports of a 100% removal efficiency using a cationic polymer [70], to date it has not been implemented on an industrial scale. It is estimated that the cost of these processes is 3.8 USD/Kg of vinasse, but the ecological impact of the emission of heavy metals and the reaction of chlorine salts with organic matter increases rather than decreases the level of toxicity. The main coagulants that are currently used in industries have metallic composition, and some environmentally friendly alternatives are being studied, including sugar polymers from some plant species, such as mesquite gum, shrimp chitosan, and some other vegetable gums [71]. Although removal percentages higher than metallic salts have been achieved by these alternatives, the process of obtaining is still expensive and uncommon.

6. Market of the bioproducts derived from agro-industrial wastes

In the last year, exploitation of agricultural waste for development of new products with a commercial value has been investigated. New related sectors have appeared in the global scene with great growth opportunities in the global market. Therefore, the specialized search engine "Web of Science," using the keyword "waste" in conjunction with the descriptive words of each item shows that the sectors with the greatest number of publications are "biocomposites and the peptides with antioxidant activity." Even though at present research in these fields remains small, it has a great potential for growth in the global market (**Figure 2**).

Globally, sectors related with products from agro-industrial residues are growing, and therefore, there is an increase in the publications showing the enormous potential to enter and take a position in the market. If we look at the size of the market for each compound, we can direct and plan new strategies aimed at the

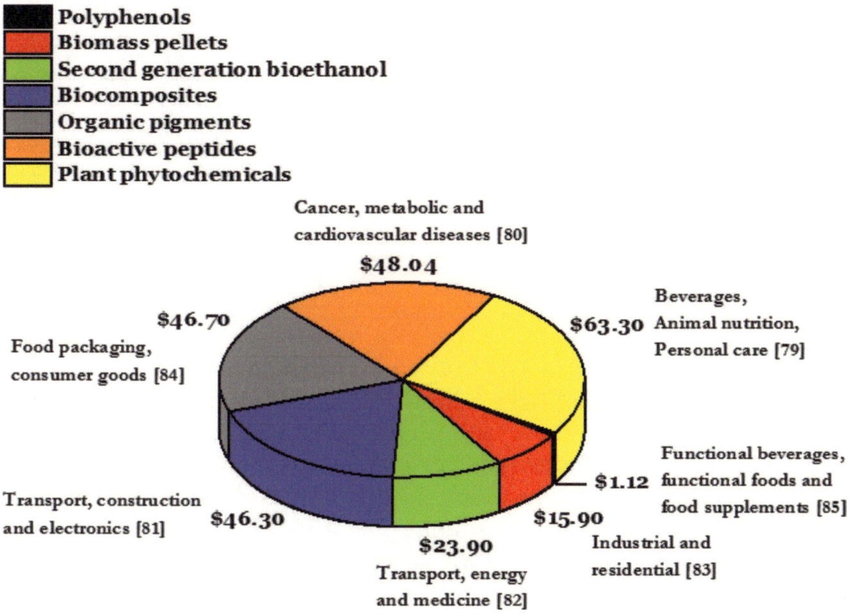

Figure 2.
Estimated market of new components obtained from waste (USD, billions) [79–85].

development of products toward the sectors with the highest economic growth and create the interest of companies looking for innovation in each of the product. The main growing market is in the phytochemicals with the extraction of carotenoids, flavonoids, and anthocyanins, among others with potential use in the alimentary industry. Valorization of residues can be achieved such as in soybean residues from pressing oil extraction, which are rich in phytochemical compounds [72]. Thus, there is a growing interest in agro-industrial residues as source of high-benefit products potentially useful as valuable constituents, flavors, and antioxidant in food and cosmetics [73]. The market of phytochemicals is becoming more competitive with the entry of pharmaceutical companies as Cargill, Hormel, and Doehler groups, which ensures a growing market. To pharmaceutical industry, bioactive peptides are considered a growing market and represent a potential solution to more efficacious disease treatment. In addition, peptides promise to combine the lower production costs and high specificity. For 2025, the proposed market growth appears to be near to USD $48.04 billion (**Figure 2**). Current trends indicate that a bright future for bioactive peptides and position them as firm candidates to the growth and innovation in pharmaceutical industry with the participation at this moment of companies as Elly Lilly and Pfizer [74], while the biocomposites are being alternatives to conventional petroleum-derived material becoming increasingly utilized in a great variety of applications [75]. An increased research is reflected in the number of publications that indicates a strong trend for applications of eco-friendly materials. Kenaf fiber has been used to reinforce polyurethane composites improving the mechanical and thermal properties [76]. The global biocomposite market is estimated to grow at USD $46.30 billion for 2025. The trends indicate that the rising awareness among people by the replacement of plastics with biodegradable and environmentally favorable alternatives allows the market growth [77]. Automotive sector is a rising market; the search for new materials that increase the safety of passengers and reduce the vehicle weight spurs market demand [78]. Biomass pellets are an emerging market with a lot of potential. The European Union was the primary market responsible for the global production and consumption of pellets to residential and district heating East Asia being predicted to become the second largest consumer [6]. Generating an advantage to closing, the circle called "the field to the hand of consumption" creates new opportunities to transform the field of an agricultural activity to agro-industrial activity centered on a circular economy. Circular economy advises the reincorporation of residues into the economy; wastes become a transient phase in an ideally perpetual utilization cycle rather than environmentally sound disposal [1].

7. Conclusions

Utilization of agro-industrial residual wastes can help to reduce them and avoid the environmental contamination, health diseases by the overcrowding, ecological damage, and other pollution-associated problems. Nowadays, there is a growing interest by researchers and industrials by the different substances and properties that exhibit the different agro-industrial residues to obtain a value-added product. Residual wastes have an enormous potential to be revalorized producing solid, liquid, and gas energies, obtaining different bioactive compounds such as polyphenols, peptides, and melanin. In addition, fibers from residues can be used to elaborate biocomposite for automotive industries mainly. New methods are proposed to obtain a value-added product from vinasses in demand not only for the pharmaceutical industries. All of these compounds and mixtures are profitable for a commercial market that grows with a high potential and application for new sectors. In that sense, the future is here.

Acknowledgements

We want thank Alejandro Carreon for kind suggestions in the early version of this manuscript.

Conflicts of interest

The authors indicate no potential conflicts of interest.

Funding

This research was supported by the project Bioturbosin Cluster (208090) of Consejo Nacional de Ciencia y Tecnología (CONACYT) and Secretaria de Energia (SENER) from Mexico.

Author details

Flora Beltrán-Ramírez, Domancar Orona-Tamayo, Ivette Cornejo-Corona, José Luz Nicacio González-Cervantes, José de Jesús Esparza-Claudio and Elizabeth Quintana-Rodríguez*
Biotecnología ambiental-Centro de Innovación Aplicada en Tecnologías Competitivas (CIATEC), Guanajuato, Mexico

*Address all correspondence to: equintana@ciatec.mx

References

[1] Velis C. Circular economy and global secondary material supply chains. Waste Management & Research [Internet]. 2015;**33**(5):389-391. DOI: 10.1177/0734242X15587641

[2] Guldhe A, Singh B, Renuka N, Singh P, Misra R, Bux F. Bioenergy: A sustainable approach for cleaner environment. In: Bauddh K, Singh B, Korstad J, editors. Phytoremediation Potential of Bioenergy Plants [Internet]. Singapore: Springer Singapore; 2017. pp. 47-62. DOI: 10.1007/978-981-10-3084-0_2

[3] Jeihanipour A, Bashiri R. Perspective of biofuels from wastes. In: Lignocellulose-Based Bioproducts. Switzerland: Springer International Publishing; 2015. pp. 37-83. DOI: 10.1007/978-3-319-14033-9

[4] Suurs RAA, Hekkert MP. Competition between first and second generation technologies: Lessons from the formation of a biofuels innovation system in the Netherlands. Energy. 2009;**34**(5):669-679. DOI: 10.1016/j.energy.2008.09.002

[5] Tauro R, García CA, Skutsch M, Masera O. The potential for sustainable biomass pellets in Mexico: An analysis of energy potential, logistic costs and market demand. Renewable and Sustainable Energy Reviews [Internet]. 2018;**82**:380-389. Available from: http://www.sciencedirect.com/science/article/pii/S1364032117312911

[6] Nunes LJR, Matias JCO, Catalão JPS. Mixed biomass pellets for thermal energy production: A review of combustion models. Applied Energy [Internet]. 2014;**127**:135-140

[7] Wang G, Luo Y, Deng J, Kuang J, Zhang Y. Pretreatment of biomass by torrefaction. Chinese Science Bulletin. 2011;**56**(14):1442-1448. DOI: 10.1007/s11434-010-4143-y

[8] Brodeur G, Yau E, Badal K, Collier J, Ramachandran KB, Ramakrishnan S. Chemical and physicochemical pretreatment of lignocellulosic biomass: A review. Enzyme Research. 2011;**2011**:17. DOI: 10.4061/2011/787532

[9] Lam PS, Lam PY, Sokhansanj S, Lim CJ, Bi XT, Stephen JD, et al. Steam explosion of oil palm residues for the production of durable pellets. Applied Energy. 2015;**141**:160-166. DOI: 10.1016/j.apenergy.2014.12.029

[10] Goetzl A. Developments in the global trade of wood pellets. Work Pap Ind US Int Trade Comm. 2015; (ID-39). Available from: https://www.usitc.gov/publications/332/wood_pellets_id-039_final_0.pdf

[11] Wang R, Shaarani SM, Godoy LC, Melikoglu M, Vergara CS, Koutinas A, et al. Bioconversion of rapeseed meal for the production of a generic microbial feedstock. Enzyme and Microbial Technology. 2010;**47**(3):77-83. DOI: 10.1016/j.enzmictec.2010.05.005

[12] Leiva-Candia DE, Pinzi S, Redel-Macías MD, Koutinas A, Webb C, Dorado MP. The potential for agro-industrial waste utilization using oleaginous yeast for the production of biodiesel. Fuel [Internet]. 2014;**123**:33-42 Available from: http://www.sciencedirect.com/science/article/pii/S0016236114000647

[13] Graça I, Lopes JM, Cerqueira HS, Ribeiro MF. Bio-oils Upgrading for Second Generation Biofuels. Industrial and Engineering Chemistry Research [Internet]. 2013;**52**(1):275-287. DOI: 10.1021/ie301714x

[14] No S-Y. Application of bio-oils from lignocellulosic biomass to transportation, heat and power generation—A review. Renewable and Sustainable Energy Reviews [Internet].

2014;**40**:1108-1125 Available from: http://www.sciencedirect.com/science/article/pii/S1364032114005796

[15] Alper K, Tekin K, Karagöz S. Pyrolysis of agricultural residues for bio-oil production. Clean Technologies and Environmental Policy [Internet]. 2015;**17**(1):211-223. DOI: 10.1007/s10098-014-0778-8

[16] Gupta A, Verma JP. Sustainable bio-ethanol production from agro-residues: A review. Renewable and Sustainable Energy Reviews [Internet]. 2015;**41**:550-567 Available from: http://www.sciencedirect.com/science/article/pii/S1364032114007084

[17] Geddes CC, Peterson JJ, Roslander C, Zacchi G, Mullinnix MT, Shanmugam KT, et al. Optimizing the saccharification of sugar cane bagasse using dilute phosphoric acid followed by fungal cellulases. Bioresource Technology [Internet]. 2010;**101**(6):1851-1857 Available from: http://www.sciencedirect.com/science/article/pii/S0960852409013200

[18] Sarkar N, Ghosh SK, Bannerjee S, Aikat K. Bioethanol production from agricultural wastes: An overview. Renewable Energy [Internet]. 2012;**37**(1):19-27 Available from: http://www.sciencedirect.com/science/article/pii/S096014811100382X

[19] Sun Y, Cheng J. Hydrolysis of lignocellulosic materials for ethanol production: A review. Bioresource Technology. 2002;**83**(1):1-11. DOI: 10.1016/S0960-8524(01)00212-7

[20] Ballesteros I, Oliva JM, Saez F, Ballesteros M. Ethanol production from lignocellulosic byproducts of olive oil extraction. Applied Biochemistry and Biotechnology. 2001;**91**(1-9):237-252. DOI: 10.1385/ABAB:91-93:1-9:237

[21] Maiti S, Sarma SJ, Brar SK, Le Bihan Y, Drogui P, Buelna G, et al. Agro-industrial wastes as feedstock for sustainable bio-production of butanol by Clostridium beijerinckii. Food and Bioproducts Processing. 2016;**98**:217-226. DOI: 10.1016/j.fbp.2016.01.002

[22] Kumar M, Gayen K. Developments in biobutanol production: New insights. Applied Energy [Internet]. 2011;**88**(6):1999-2012 Available from: http://www.sciencedirect.com/science/article/pii/S0306261910005751

[23] Qureshi N, Saha BC, Hector RE, Hughes SR, Cotta MA, Qureshi N. Butanol production from wheat straw by simultaneous saccharification and fermentation using Clostridium beijerinckii: I. Batch fermentation. Biomass and Bioenergy. 2008;**32**(2):168-175. DOI: 10.1016/j.biombioe.2007.07.004

[24] Al-Shorgani NKN, Kalil MS, Yusoff WMW. Biobutanol production from rice bran and de-oiled rice bran by *Clostridium saccharoperbutylacetonicum* N1-4. Bioprocess and Biosystems Engineering. 2012;**35**(5):817-826. DOI: 10.1007/s00449-011-0664-2

[25] Deublein D, Steinhauser A. Biogas from Waste and Renewable Resources: An Introduction. Germany: John Wiley & Sons; 2011. DOI: 10.1002/9783527621705

[26] Solarte-Toro JC, Chacón-Pérez Y, Cardona-Alzate CA. Evaluation of biogas and syngas as energy vectors for heat and power generation using lignocellulosic biomass as raw material. Electronic Journal of Biotechnology. 2018;**33**:56-62. DOI: 10.1016/j.ejbt.2018.03.005

[27] Wellinger A, Murphy JD, Baxter D. The Biogas Handbook: Science, Production and Applications. Woodhead United Kingdom: Elsevier; 2013. Available from: http://www.iea-biogas.net

[28] Zheng Y, Zhao J, Xu F, Li Y. Pretreatment of lignocellulosic biomass

for enhanced biogas production. Progress in Energy and Combustion Science. 2014;**42**:35-53. DOI: 10.1016/j.pecs.2014.01.001

[29] Song Z, Liu X, Yan Z, Yuan Y, Liao Y. Comparison of seven chemical pretreatments of corn straw for improving methane yield by anaerobic digestion. PLoS One. 2014;**9**(4):e93801DOI. DOI: 10.1371/journal.pone.0093801

[30] Ignat I, Volf I, Popa VI. A critical review of methods for characterisation of polyphenolic compounds in fruits and vegetables. Food Chemistry. 2011;**126**(4):1821-1835. DOI: 10.1016/j.foodchem.2010.12.026

[31] Ayala-Zavala JF et al. Antioxidant enrichment and antimicrobial protection of fresh-cut fruits using their own byproducts: Looking for integral exploitation. Journal of Food Science, Wiley Online Library. 2010;**75**(8):R175-R181. DOI: 10.1111/j.1750-3841.2010.01792.x

[32] Joshi VK, Kumar A, Kumar V. Antimicrobial, antioxidant and phyto-chemicals from fruit and vegetable wastes: A review. International Journal of Food and Fermentation Technology. 2012;**2**(2):123 Available from: http://ndpublisher.in/admin/issues/ijfftv2n2c.pdf

[33] Yusuf M. Agro-industrial waste materials and their recycled value-added applications. Handbook of Ecomaterials. 2017:1-11. DOI: 10.1007/978-3-319-48281-1_48-1

[34] Pujol D, Liu C, Gominho J, Olivella MÀ, Fiol N, Villaescusa I, et al. The chemical composition of exhausted coffee waste. Industrial Crops and Products. 2013;**50**:423-429. DOI: 10.1016/j.indcrop.2013.07.056

[35] Fu R, Zhang Y, Guo Y, Liu F, Chen F. Determination of phenolic contents and antioxidant activities of extracts of Jatropha curcas L. seed shell, a by-product, a new source of natural antioxidant. Industrial Crops and Products. 2014;**58**:265-270. DOI: 10.1016/j.indcrop.2014.04.031

[36] Abbas M, Ali A, Arshad M, Atta A, Mehmood Z, Tahir IM, et al. Mutagenicity, cytotoxic and antioxidant activities of Ricinus communis different parts. Chemistry Central Journal. 2018;**12**(1):3. DOI: 10.1186/s13065-018-0370-0

[37] Mishra B, Varjani S, Varma GKS. Agro-industrial by-products in the synthesis of food grade microbial pigments: An eco-friendly alternative. In: Green Bio-processes. Singapore: Springer; 2019. pp. 245-265. DOI: 10.1007/978-981-13-3263-0_13

[38] Kantifedaki A, Kachrimanidou V, Mallouchos A, Papanikolaou S, Koutinas AA. Orange processing waste valorisation for the production of bio-based pigments using the fungal strains *Monascus purpureus* and *Penicillium purpurogenum*. Journal of Cleaner Production. 2018;**185**:882-890. DOI: 10.1016/j.jclepro.2018.03.032

[39] Rodrigues DB, Flores ÉMM, Barin JS, Mercadante AZ, Jacob-Lopes E, Zepka LQ. Production of carotenoids from microalgae cultivated using agroindustrial wastes. Food Research International. 2014;**65**:144-148. DOI: 10.1016/j.foodres.2014.06.037

[40] Tarangini K, Mishra S. Production of melanin by soil microbial isolate on fruit waste extract: Two step optimization of key parameters. Biotechnology Reports. 2014;**4**:139-146. Available from: https://pdfs.semanticscholar.org/1832/5493d429f8d8986c892b637f9f6aa28065ec.pdf

[41] Zou Y, Hu W, Ma K, Tian M. Fermentative production of melanin by the fungus *Auricularia auricula* using wheat bran extract as major nutrient

source. Food Science and Technology Research. 2017;**23**(1):23-29. DOI: 10.1111/jfpp.12909

[42] Tarangini K, Mishra S. Production, characterization and analysis of melanin from isolated marine Pseudomonas sp. using vegetable waste. Research Journal of Engineering Sciences. 2013;**2278**:9472 Available from: https://pdfs.semanticscholar.org/1832/5493d429f8d8986c892b637f9f6aa28065ec.pdf

[43] Keles Y, Özdemir Ö. Extraction, purification, antioxidant properties and stability conditions of phytomelanin pigment on the sunflower seeds. International Journal of Secondary Metabolite. 2018;**5**(2):140-148. DOI: 10.21448/ijsm.377470

[44] Kartushina YN, Nefedieva EE, Sevriukova GA, Gracheva NV, Zheltobryukhov VF. Technological desition of extraction of melanin from the waste of production of sunflower-seed oil. In: IOP Conference Series: Earth and Environmental Science. IOP Publishing; 2017. p. 12014. DOI: 10.1088/1755-1315/66/1/012014

[45] Lan J, Wang J, Wang F. Study on the antioxidant capacity of the melanin from walnut shell and walnut epicarp. Science and Technology of Food Industry. 2012;**15**:19. Available from: http://en.cnki.com.cn/Article_en/CJFDTOTAL-SPKJ201215019.htm

[46] Orona-Tamayo D, Valverde ME, Paredes-López O. Bioactive peptides from selected Latin American food crops – A nutraceutical and molecular approach. Critical Reviews in Food Science and Nutrition. 2018:1-25. DOI: 10.1080/10408398.2018.1434480. https://www.tandfonline.com/doi/full/10.1080/10408398.2018.1434480

[47] Lemes A, Sala L, Ores J, Braga A, Egea M, Fernandes K. A review of the latest advances in encrypted bioactive peptides from protein-rich waste. International Journal of Molecular Sciences. 2016;**17**(6):950. DOI: 10.3390/ijms17060950

[48] León-Villanueva A, Huerta-Ocampo JA, Barrera-Pacheco A, Medina-Godoy S, de la Rosa APB. Proteomic analysis of non-toxic Jatropha curcas byproduct cake: Fractionation and identification of the major components. Industrial Crops and Products. 2018;**111**:694-704. DOI: 10.1016/j.indcrop.2017.11.046

[49] Devappa RK, Makkar HPS, Becker K. Nutritional, biochemical, and pharmaceutical potential of proteins and peptides from Jatropha. Journal of Agricultural and Food Chemistry. 2010;**58**(11):6543-6555. DOI: 10.1021/jf100003z

[50] Souza PFN, Vasconcelos IM, Silva FDA, Moreno FB, Monteiro-Moreira ACO, Alencar LMR, et al. A 2S albumin from the seed cake of *Ricinus communis* inhibits trypsin and has strong antibacterial activity against human pathogenic bacteria. Journal of Natural Products. 2016;**79**(10):2423-2431. DOI: 10.1021/acs.jnatprod.5b01096

[51] Faruk O, Bledzki AK, Fink HP, Sain M. Biocomposites reinforced with natural fibers: 2000-2010. Progress in Polymer Science. 2012;**37**(11):1552-1596. DOI: 10.1016/j.progpolymsci.2012.04.003

[52] Zuccarello B, Scaffaro R. Experimental analysis and micromechanical models of high performance renewable agave reinforced biocomposites. Composites. Part B, Engineering. 2017;**119**:141-152. DOI: 10.1016/j.compositesb.2017.03.056

[53] Sengupta A, Pattnaik S, Sutar MK. Biocomposites: An overview. International Journal of Engineering, Technology, Science and Research IJETSR. 2017;**4**:2394-3386. Available from: www.ijetsr.com

[54] Xie Y, Xiao Z, Gruneberg T, Militz H, Hill CAS, Steuernagel L, et al. Effects of chemical modification of wood particles with glutaraldehyde and 1,3-dimethylol-4,5-dihydroxyethyleneurea on properties of the resulting polypropylene composites. Composites Science and Technology. 2010;**70**(13):2003-2011. DOI: 10.1016/j.compscitech.2010.07.024

[55] Kaewkuk S, Sutapun W, Jarukumjorn K. Effects of interfacial modification and fiber content on physical properties of sisal fiber/polypropylene composites. Composites. Part B, Engineering. 2013;**45**(1):544-549. DOI: 10.1016/j.compositesb.2012.07.036

[56] Sood M, Dwivedi G. Effect of fiber treatment on flexural properties of natural fiber reinforced composites : A review. Egyptian Journal of Petroleum. 2018;**27**:775-783. DOI: 10.1016/j.ejpe.2017.11.005

[57] Koronis G, Silva A, Fontul M. Green composites: A review of adequate materials for automotive applications. Composites. Part B, Engineering. 2013;**44**(1):120-127. DOI: 10.1016/j.compositesb.2012.07.004

[58] Pracella M, Chionna D, Anguillesi I, Kulinski Z, Piorkowska E. Functionalization, compatibilization and properties of polypropylene composites with Hemp fibres. Composites Science and Technology. 2006;**66**(13):2218-2230. DOI: 10.1016/j.compscitech.2005.12.006

[59] Vilaseca F, Valadez-Gonzalez A, Herrera-Franco PJ, Pèlach MÀ, López JP, Mutjé P. Biocomposites from abaca strands and polypropylene. Part I: Evaluation of the tensile properties. Bioresource Technology [Internet]. 2010;**101**(1):387-395 Available from: http://www.sciencedirect.com/science/article/pii/S0960852409009638

[60] Bledzki AK, Franciszczak P, Osman Z, Elbadawi M. Polypropylene biocomposites reinforced with softwood, abaca, jute, and kenaf fibers. Industrial Crops and Products. 2015;**70**:91-99. DOI: 10.1016/j.indcrop.2015.03.013

[61] Khalil HPSA, Aprilia NAS, Bhat AH, Jawaid M, Paridah MT, Rudi D. A Jatropha biomass as renewable materials for biocomposites and its applications. Renewable and Sustainable Energy Reviews. 2013;**22**:667-685. DOI: 10.1016/j.rser.2012.12.036

[62] Zuccarello B, Zingales M. Toward high performance renewable agave reinforced biocomposites: Optimization of fiber performance and fiber-matrix adhesion analysis. Composites. Part B, Engineering. 2017;**122**:109-120. DOI: 10.1016/j.compositesb.2017.04.011

[63] Vinayaka DL, Guna V, Madhavi D, Arpitha M, Reddy N. Ricinus communis plant residues as a source for natural cellulose fibers potentially exploitable in polymer composites. Industrial Crops and Products. 2017;**100**:126-131. DOI: 10.1016/j.indcrop.2017.02.019

[64] España-Gamboa E, Mijangos-Cortes J, Barahona-Perez L, Dominguez-Maldonado J, Hernández-Zarate G, Alzate-Gaviria L. Vinasses: Characterization and treatments. Waste Management & Research. 2011;**29**(12):1235-1250. DOI: 10.1177/0734242X10387313

[65] Martínez-Gutiérrez GA, Ortiz-Hernández YD, Aquino-Bolaños T, Bautista-Cruz A, López-Cruz JY. Properties of *Agave angustifolia* Haw. bagasse before and after its composting. Comunicata Scientiae. 2015;**6**(4):418-429

[66] Moraes BS, Zaiat M, Bonomi A. Anaerobic digestion of vinasse from sugarcane ethanol production in Brazil: Challenges and perspectives. Renewable

and Sustainable Energy Reviews. 2015;**44**:888-903. DOI: 10.1016/j.rser.2015.01.023

[67] Christofoletti CA, Escher JP, Correia JE, Marinho JFU, Fontanetti CS. Sugarcane vinasse: Environmental implications of its use. Waste Management. 2013;**33**(12):2752-2761. DOI: 10.1016/j.wasman.2013.09.005

[68] López-López A, Davila-Vazquez G, León-Becerril E, Villegas-García E, Gallardo-Valdez J. Tequila vinasses: Generation and full scale treatment processes. Reviews in Environmental Science and Bio/Technology. 2010;**9**(2):109-116. DOI: 10.1007/s11157-010-9204-9

[69] Rodríguez-Félix E, Contreras-Ramos SM, Davila-Vazquez G, Rodríguez-Campos J, Marino-Marmolejo EN. Identification and quantification of volatile compounds found in vinasses from two different processes of tequila production. Energies. 2018;**11**(3):490. DOI: 10.3390/en11030490

[70] Carpinteyro-Urban S, Vaca M, Torres LG. Can vegetal biopolymers work as coagulant–flocculant aids in the treatment of high-load cosmetic industrial wastewaters? Water, Air. Soil Pollution. 2012;**223**(8):4925-4936. DOI: 10.1007/s11270-012-1247-9

[71] Torres LG, Carpinteyro-Urban SL. Use of Prosopis laevigata seed gum and Opuntia ficus-indica mucilage for the treatment of municipal wastewaters by coagulation-flocculation. Natural Resources Research. 2012;**3**(2):35. DOI: 10.4236/nr.2012.32006

[72] Alvarez MV, Cabred S, Ramirez CL, Fanovich MA. Valorization of an agroindustrial soybean residue by supercritical fluid extraction of phytochemical compounds. Journal of Supercritical Fluids. 2019;**143**:90-96. DOI: 10.1016/j.supflu.2018.07.012

[73] Spatafora C, Tringali C. Valorization of vegetable waste: Identification of bioactive compounds and their chemo-enzymatic optimization. The Open Agriculture Journal. 2012;**6**(1). DOI: 10.2174/1874331501206010009

[74] Uhlig T, Kyprianou T, Martinelli FG, Oppici CA, Heiligers D, Hills D, et al. The emergence of peptides in the pharmaceutical business: From exploration to exploitation. EuPA Open Proteomics. 2014;**4**:58-69. DOI: 10.1016/j.euprot.2014.05.003

[75] Väisänen T, Haapala A, Lappalainen R, Tomppo L. Utilization of agricultural and forest industry waste and residues in natural fiber-polymer composites: A review. Waste Management. 2016;**54**:62-73. DOI: 10.1016/j.wasman.2016.04.037

[76] El-Shekeil YA, Sapuan SM, Abdan K, Zainudin ES. Influence of fiber content on the mechanical and thermal properties of Kenaf fiber reinforced thermoplastic polyurethane composites. Materials and Design. 2012;**40**:299-303. DOI: 10.1016/j.matdes.2012.04.003

[77] Ariadurai S. Bio-composites: Current status and future trends. Research Gate. Available from: https://www.researchgate.net/profile/Samuel_Ariadurai/publication/256308472_Bio-Composites_Current_Status_and_Future_Trends/links/02e7e5224a0558244d000000.pdf

[78] Gurunathan T, Mohanty S, Nayak SK. A review of the recent developments in biocomposites based on natural fibres and their application perspectives. Composites. Part A, Applied Science and Manufacturing. 2015;**77**:1-25. DOI: 10.1016/j.compositesa.2015.06.007

[79] Coherent marketing insights. Global Plant Extracts Market to Surpass US$ 63.30 Billion by 2025. 2018. Available from: https://www.marketsandmarkets.com

[80] Grand View Research. Peptide Therapeutics Market Size Worth $ 48 . 04 Billion By 2025. 2018;1-7. Available from: https://www.grandviewresearch.com/press-release/global-peptide-therapeutics-market

[81] Grand View Research. Biocomposites Market Size Worth $46.3 Billion By 2025 | CAGR 12.5%. 2018. Available from: https://www.grandviewresearch.com/press-release/global-biocomposites-market

[82] Allied Market Research. Second Generation Biofuels Market Size, Share and Trends. 2018. Available from: https://www.alliedmarketresearch.com/second-generation-biofuels-market

[83] Zion Market Research. Global Biomass Pellets Market worth USD 15.9 Billion by 2022. 2017. Available from: https://www.zionmarketresearch.com/news/biomass-pellets-market

[84] Markets and markets. Organic Pigments Market by Application & Type—Global Forecast 2023 | MarketsandMarketsTM. 2018. Available from: https://www.marketsandmarkets.com/Market-Reports/organic-pigments-market-1076.html?gclid=EAIaIQobChMIgMa1l4T

[85] Allied market research. Polyphenol Market Size, Share and Trends | Industry Analysis, 2022. 2018. Available from: https://www.alliedmarketresearch.com/polyphenol-market

Chapter 6

Bio-hydrogen and Methane Production from Lignocellulosic Materials

Apilak Salakkam, Pensri Plangklang, Sureewan Sittijunda, Mallika Boonmee Kongkeitkajorn, Siriporn Lunprom and Alissara Reungsang

Abstract

This chapter covers the information on bio-hydrogen and methane production from lignocellulosic materials. Pretreatment methods of lignocellulosic materials and the factors affecting bio-hydrogen production, both dark- and photo-fermentation, and methane production are addressed. Last but not least, the processes for bio-hydrogen and methane production from lignocellulosic materials are discussed.

Keywords: anaerobic digestion, pretreatment, biomass, bioconversion, fermentation process, bio-hydrogen, methane

1. Introduction

The need for energy has continuously been a major issue in human society. Energy use per capita has been increasing at an average rate of 21.5 kg of oil equivalent annually since the year 2000 (value calculated from [1]). Increase in energy demands leads to the search for alternative sources for energy production. Biomass, as the fourth largest energy source after coal, oil, and natural gas, is a very promising resource for energy production due to its renewability and versatility [2]. Biomass is biologically originated materials or simply any materials that are not fossilized. Supplies of biomass could be from forestry, agriculture, and wastes. They could be used directly to produce energy by burning or could be refined to produce biofuels in the form of solid, liquid, or gas [3].

Lignocellulosic biomass is the biomass with the structure that is composed of lignin, hemicellulose, and cellulose. They can be divided into woody and non-woody biomass. Woody biomass can be further categorized into hardwoods and softwoods, which differ in their reproduction; angiosperm for hardwoods and gymnosperm for softwoods. Examples of hardwoods include beech, mahogany, maple, and teak, while softwoods are cedar, pine, juniper, and spruce. Non-woody biomass are those of agricultural residues, grass family (Poaceae or Gramineae), and non-woody fibers such as cotton fiber [4, 5].

Apart from using biomass in co-firing with fossil fuels, gasification, and pyrolysis, its use in fermentation technology for liquid and gaseous biofuels production is also applicable and widely studied. The hemicellulose and cellulose structures (so called holocellulose) in lignocellulosic biomass contain sugar monomers that could be utilized by microorganisms and converted to various biofuels via biological pathways.

Cellulose is considered a major composition of lignocellulosic biomass. It is a homopolymer containing glucose as the only monomer. Glucose molecules in cellulose are linked by β-1,4-glycosidic bonds (**Figure 1**). Cellulose chain is also known as β-1,4-glucan. The chains are packed into tiny and extremely long structure called microfibrils. These microfibrils are packed into lattices, which make most part of the cellulose fibers inaccessible by enzymes [6–8].

Hemicellulose is also the composition that consists of sugars that can be utilized by microorganisms. It is a heteropolymer of pentoses (xylose, arabinose), hexoses (mannose, glucose, galactose), and uronic acids (4-O-methylglucuronic acid, galacturonic acid). Due to its heterogeneity, different structures of hemicellulose are found in different types of biomass [9, 10].

In hardwoods, glucuronoxylan is the major hemicellulose. Its backbone consists of xylose connected by β-1,4-glycosidic linkage, with some acetylation at C2 and C3 of xylose molecules. In addition, side chains of 4-O-methylglucuronic acid are found attached to xylose with α-1,2-linkage. Main hemicellulose in softwoods is galactoglucomannan. As the name suggests, galactoglucomannan has mannose and glucose as the backbone, with galactose and acetyl group as the side chains. For grasses (including cereals), glucuronoarabinoxylan, wherein xylose is the backbone, is the major hemicellulose [9, 11]. Structures of main hemicellulose in lignocellulosic biomass are illustrated in **Figure 2**.

Some other hemicellulose, which can be found in multiple sources, include xyloglucan which could be found in all hardwoods, softwoods, and grasses, arabinoglucuronoxylan in grasses and softwood. In addition, glucomannan is found as a minor component in softwoods and hardwoods.

The last major component of lignocellulosic biomass is lignin. Lignin is the only non-sugar component of the biomass. It is the second most abundant biopolymer besides cellulose. It is an amorphous polymer with structures that vary among different types of biomass and environmental conditions. Primarily, lignin consisted of three phenylpropane units of p-hydroxyphenyl (H), guaiacyl (G), and syringyl (S), which are originated from aromatic alcohols, p-coumaryl, coniferyl, and sinapyl alcohols [12].

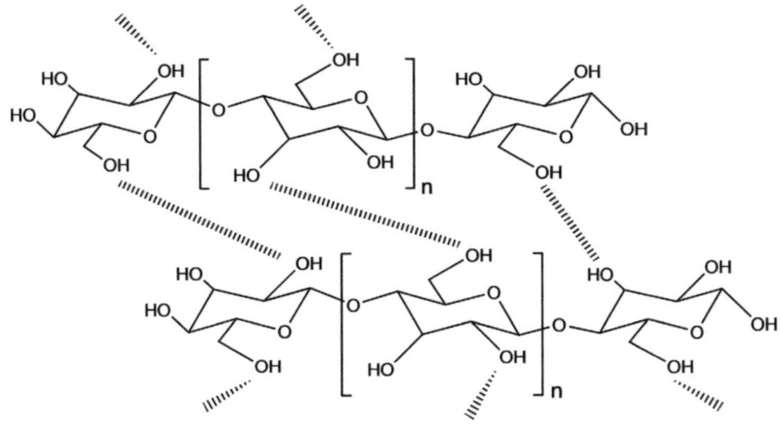

Figure 1.
Structure of cellulose.

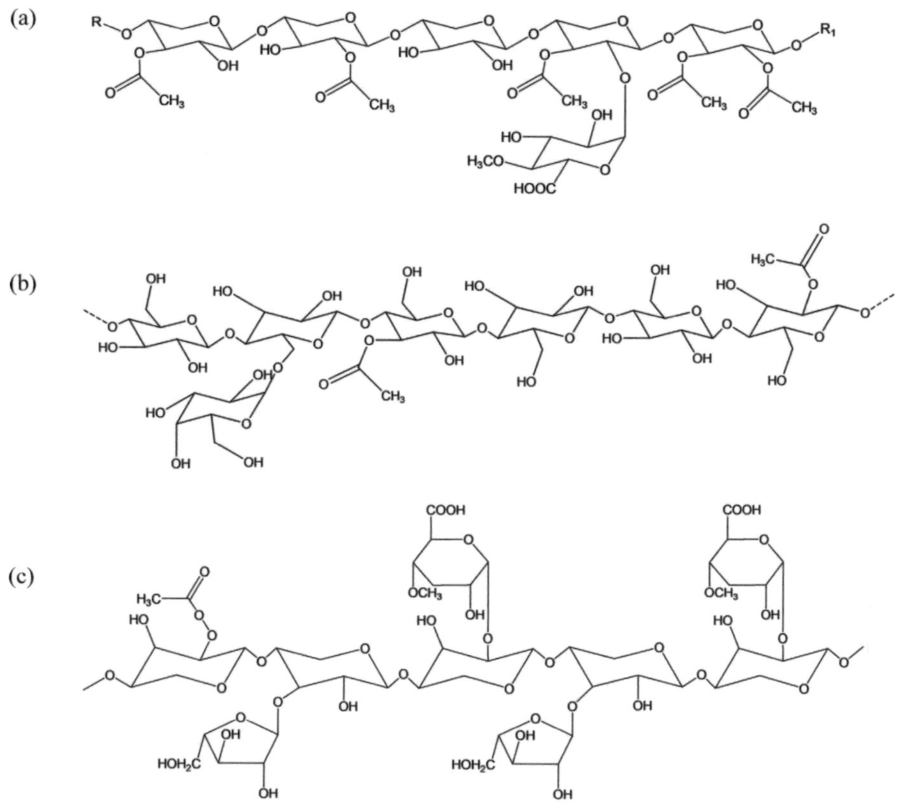

Figure 2.
Structures of main hemicellulose in hardwoods, soft woods and grasses. (a) Glucorunoxylan, (b) galactoglucomannan, (c) glucuronoarabinoxylan.

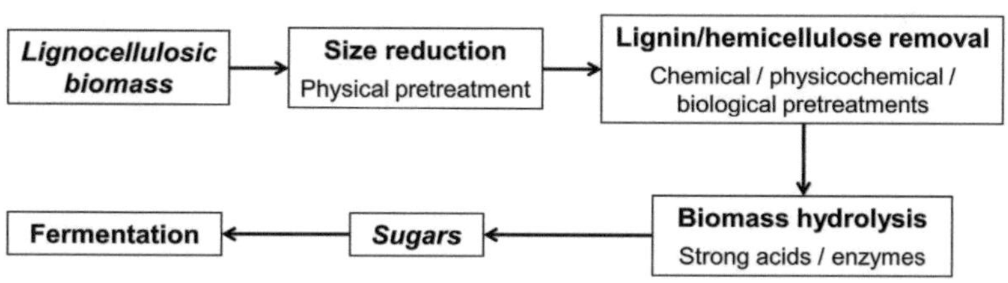

Figure 3.
Basic flow diagram for the use of lignocellulosic biomass in biofuels production through fermentation route.

Although the sugar monomers in holocellulose part of the biomass are of interest for use in biofuel production via biological pathways, breaking down the structure to obtain the monomers is not a simple task. All three components of lignocellulose are incorporated into complex structures and recalcitrant to hydrolysis. Not only the cellulose itself has a strong crystalline structure, its microfibrils are packed and interconnected with hemicellulose. In addition, lignin that fills the void of the structure adds additional strength, increases the hydrophobicity of the wall, and hence prevents the action of hydrolytic enzymes [13, 14].

In order to utilize lignocellulosic biomass in production of biofuels via biological pathways, its tough structures have to be loosen, and hydrolysis of holocellulose needs to be achieved to release sugars for microbial usage. Block diagram in **Figure 3** shows generalized scheme for handling and processing of lignocellulosic biomass when applied in biofuel production by microorganisms through fermentation process.

2. Pretreatment of lignocellulosic material

Lignocellulosic biomass is abundantly available, relatively low-cost, and is a good feedstock for the production of biofuels due to their compositions (cellulose, hemicellulose, and lignin). The natural microorganisms cannot directly ferment lignocellulosic biomass into biofuels. The pretreatment step is required to overcome the recalcitrance attributed to the structural characteristic of lignocellulosic biomass and hydrolyze the lignocellulose biomass into fermentation sugars. Various pretreatment technologies have been proposed, challenging the complexity of biomass structure and attempting to recover high fermentable sugars. The pretreatment methods must meet the following requirements: (1) increase the sugar production or ability to afterward form sugar by enzymatic hydrolysis, (2) minimize the formation of inhibitors that affect the hydrolysis and fermentation process, (3) avoid the loss of carbohydrates, and (4) be cost-effective. The present section summarizes the performance of various pretreatment technologies, including physical, chemical, physicochemical, and biological processes. Furthermore, the advantages and disadvantages of different pretreatment technologies are also included.

2.1 Physical pretreatment

Physical pretreatment involves an increase in the accessible surface area of lignocellulosic materials to enzymes by breaking down the particle size or disrupting their crystalline structures. The physical pretreatment methods such as chipping, milling, and grinding are applied to pretreat several lignocellulosic materials [15]. Chipping and grinding are used to reduce a huge lignocellulosic material into small pieces. Thus, milling is required to mill lignocellulosic material into fine particles. Among these physical methods, milling can significantly reduce the degree of crystallinity and particle size and consequently improve their enzymatic hydrolysis [16]. The energy requirement for physical pretreatment methods depends on the particle size and the reduction of crystallinity in lignocellulosic material. In fact, the required energy is higher than the theoretical energy content available in the biomass [15]. As aforementioned, these methods cannot be used in an industrial scale process due to its cost.

Microwave irradiation is another physical pretreatment method. It is a heating method which directly applies an electromagnetic field to the molecular structure. Microwaves are nonionizing electromagnetic radiation with the wavelengths ranging from 1 mm to 1 m. The electromagnetic spectrums are located between 300 and 300,000 MHz. The application of microwave pretreatment causes swelling and fragmentation of lignocellulosic biomass. The study of Shahzadi et al. [17] indicates that the use of microwave irradiation can enhance the digestibility of lignocellulosic material. In order to enhance the hydrolysis efficiency, microwave pretreatment assisted with catalysts such as acid and alkaline are applied [18]. The advantages and disadvantages of physical pretreatment method are tabulated in **Table 1**.

2.2 Chemical pretreatment

Acid, alkaline, ionic liquid, and organic solvent (organosolv) are used as catalysts in the chemical pretreatment methods. Since 1819, acids including sulfuric and hydrochloric are applied to pretreat lignocellulosic materials [19]. After the discovery, various concentrated and diluted acids have been used to pretreat various lignocellulosic materials [20, 21]. The concentrated acid pretreatments can degrade

Pretreatment method	Effects	Advantages	Disadvantages
Physical			
Chipping, grinding, milling	Reduce the particle size and disrupt the crystallinity	Control of final particle size, easy handling, less water consumption	High energy consumption
Microwave	Swelling and fragmentation of lignocellulosic material	Fast heat transfer, short reaction time, energy-efficient	Low penetration of radiation in bulk products, the distribution of microwave power around of material due to nonhomogeneous material
Chemical			
Acid	Lignin cellulose and hemicellulose fractionate	Enzymatic hydrolysis is sometimes not required as the acid itself may hydrolyze the biomass to fermentable sugars	Corrosive and toxic, formation of inhibitors as by-products
Alkaline	Lignin and hemicellulose removal	Reduce the absorption of cellulose due to efficient lignin removal, low cost	Generates inhibitors, long residence time required
Ionic liquid	Cellulose precipitation and lignin removal	Working under mild reaction condition, low vapor pressure	High cost, complexity of purification and synthesis
Organosolv	Lignin removal and hemicellulose fractionate	Formation of a high purity of lignin	High capital cost, need to separate solvent, need washing step
Physicochemical			
Steam explosion	Particle size reduction, partial hydrolysis of hemicellulose, lignin removal	Less water uses, no chemical uses, low environmental impacts	It has a high equipment cost
Liquid hot water	Partial hydrolysis of hemicellulose, lignin removal	Does not require washing, chemical recovery, or detoxification steps	High water consumption and energy input
AFEX	Decreases the crystallinity and lignin removal	Low formation of by-products	Not suitable for lignocellulosic biomass with high lignin content
SPORL	Lignin removal and hemicellulose fractionate	Low formation of inhibitors, energy-efficient, reduces the absorption by sulfonation of cellulose	High cost of chemical recovery
Biological			
Microorganisms and enzymes	Lignin, hemicellulose, and cellulose degradation	Selective degradation of lignin, hemicellulose, and cellulose, environmentally friendly	Long pretreatment time, the hydrolysis rate is low

Table 1.
Summary of advantages and disadvantages of each pretreatment methods [22–24].

cellulose and produce a high concentration of inhibitors, such as furfural and 5-hydroxymethylfurfural (5-HMF). In addition, the utilization of concentrated acid causes corrosion of equipment, making the process less attractive [21]. Dilute acid is an attractive method due to its ability to hydrolyze both hemicellulose and cellulose. As results, pentose sugars (xylose and arabinose) and hexose (glucose) sugar are obtained in the hydrolysate. Moreover, this process minimizes the inhibitor formation compared with concentrated acid pretreatment. However, both concentrated and diluted acids slightly degrade lignin.

Alkaline pretreatment is the most commonly used to degrade lignin in lignocellulosic material. Alkaline reagents, such as sodium hydroxide (NaOH), potassium hydroxide (KOH), calcium hydroxide ($Ca(OH)_2$), aqueous ammonia (NH_4OH), and oxidative alkaline, are mainly used to cleave the ester linkages in lignin and hemicellulose structures. The cleavages of these linkages significantly enhance the solubilization of lignin and hemicellulose, resulting in a higher cellulose hydrolysis to fermentable sugar by microorganisms or enzymatic hydrolysis.

Ionic liquids (ILs) are salts composed of cations and anions. These liquids have melting point lower than 100°C and low vapor pressure [25]. Anions and cations in ILs form hydrogen bonds with cellulose hydroxyl groups, resulting in a cellulose precipitation. In addition, lignin can be dissolved in the ILs [25]. This reaction occurs in mild conditions with the ease of cellulose recovery, as well as the ILs, with no toxic or odor emission. However, the utilization of ILs to pretreat lignocellulosic materials is not favorable due to its cost. In comparison to other chemical pretreatments, ILs have the advantages of low toxicity, high solvation power, low volatility, thermal stability, as well as inflammability.

In the organosolv process, organic solvents are mainly used to cleave the linkage of lignin and hemicellulose which can increase the pore volume and accessible surface area of cellulose. The resulting lignin is dissolved in the organic solvent phase, while cellulose is recovered as the solid. Many organic solvents such as ethanol, methanol, acetone, organic acids, and ethylene glycol have been utilized to pretreat various lignocellulosic materials. Among these, ethanol is the most favorable solvent due to its low toxicity and its ease of recovery. This process can occur in the presence or absence of catalysts (acid or base) [26]. Comparing with other chemical pretreatments, organosolv process has many advantages such as easy to recover solvent by distillation, low environmental impact, and recovery of high-quality lignin as by-product. Contrastingly, high price of organic solvent and potential hazard of handling large volume of organic solvents limit the utilization of organosolv process. The overall advantages and disadvantages of chemical pretreatments are shown in **Table 1**.

2.3 Physicochemical pretreatment

Physicochemical pretreatment is a combination between physical and chemical pretreatments, which aims to enhance lignin removal and increase the hydrolysis efficiency. Several successful physicochemical pretreatments, such as steam explosion, liquid hot water, wet oxidation, ammonia-based, and sulfite pretreatment (SPORL), are applied to various lignocellulosic materials.

Steam explosion is a combined method between thermo-mechano-chemical treatments. In this process, biomass is exposed to a high pressure (0.69–4.83 MPa) with a saturated steam at a high temperature (160–260°C) for a few seconds [21, 27]. The steam penetrates into the biomass and swells the cell wall of the fibers before the explosion and partial hydrolysis. During pretreatment, the hydrolysis of hemicellulose into hexose and pentose sugars is accomplished by the action of acetic acid produced from the acetyl groups of hemicellulose. This process is called "autohydrolysis." The efficiency of steam explosion can be enhanced by adding the catalyst such as sulfuric acid (H_2SO_4), SO_2, or CO_2. Among these catalysts, acid is the best in terms of sugar recovery, minimization of the inhibition compound formation, and enzymatic hydrolysis improvement [21]. Liquid hot water pretreatment process is quite similar to steam explosion pretreatment, but it uses water instead of steam. This leads to less formation of inhibitors at the high temperatures.

In the ammonia-based or ammonia fiber explosion (AFEX) process, the lignocellulosic biomass is subjected to liquid ammonia at a high pressure (250–300 psi) and a temperature around 60–100°C for a few minutes. After that, the pressure is immediately released [28]. Liquid ammonia can cause the swelling of lignocellulose structure, resulting in an increase in the enzymatic hydrolysis efficiency. The immediate release of the pressure causes the physical disruption in the crystalline cellulose, resulting in a decrease in the crystallinity of lignocellulosic biomass. However, the lignin and hemicellulose degradation efficiency is low. AFEX process has advantages such as mild reaction temperature and low formation of inhibitors.

SPORL pretreatment process consists of two steps. First, the lignocellulosic materials are treated with magnesium sulfite or calcium sulfite in order to remove the lignin and hemicellulose fractions. Second, the mechanical disk miller is used to reduce the particle size of pretreated lignocellulosic material. This method is efficient to pretreat various lignocellulosic materials [21]. The amounts of HMF and furfural generated from SPORL pretreatment are less than those obtained using acid pretreatment. This is attributed to the fact that at the same acid charge, higher amount of bisulfite leads to higher pH which reduces the decomposition of sugar to HMF and furfural [21].

2.4 Biological pretreatment

In the biological pretreatment, microorganisms and enzymes are the key points used to pretreat lignocellulosic materials before enzymatic hydrolysis [22, 28]. Main biological process is delignification and saccharification process. Microorganisms, such as brown, white, and soft rot fungi, have been used to degrade lignocellulosic materials. White and soft rot fungi mainly degrade lignin and hemicellulose while brown rot fungi are used to degrade cellulose [22, 28]. White rot fungi such as *Cyathus stercoreus*, *Phanerochaete chrysosporium*, *Ceriporia lacerata*, *Ceriporiopsis subvermispora*, *Pycnoporus cinnabarinus*, and *Pleurotus ostreatus* are frequently applied to degrade lignin because these species contain lignin degradation enzymes, including peroxidase and laccase [22, 28]. Also, Basidiomycetes species, such as *Bjerkandera adusta*, *Irpex lacteus*, *Fomes fomentarius*, and *Trametes versicolor* are studied for breaking down lignocellulosic materials [11, 12]. Recently, cellulose hydrolyzing bacteria such as Clostridia and Actinomycetes are widely used to pretreat lignocellulosic materials. Clostridia and Actinomycetes grow and degrade lignocellulose under anaerobic and aerobic conditions, respectively [29]. Clostridia have an extracellular complex enzyme system called "cellulosome" that can degrade lignocellulosic materials. This system contains various enzymes, such as endoglucanases, exoglucanases, hemicellulases, chitinases, pectin lyases, and lichenases [30].

As for enzymes used in biological pretreatment, both commercial and extracted enzymes from microbes are used. Commercial cellulase and xylanase are commonly used to degrade lignocellulosic materials such as sugarcane bagasse [31], rice straw [32], napier grass [33], etc. Extracted lignin degradation enzymes, including lignin peroxidase, manganese peroxidase, and laccase, from white rot fungi, are also used to degrade lignin from lignocellulosic materials [28]. The study of Taniguchi et al. [34] found that pretreating rice straw with *Pleurotus ostreatus* enhanced the degradation of lignin and hemicellulose to 41 and 48% degradation efficiency, respectively. The lignin and hemicellulose degradation by *Pleurotus ostreatus* occurs through the action of peroxidase and laccase. Biological pretreatment is environmentally friendly as no chemicals and lower energy are used compared with other pretreatment methods. The advantages and disadvantages of biological pretreatment methods are given in **Table 1**.

Currently, the combined physical, chemical, and biological pretreatment process is investigated for enhancing the degradation efficiency [21]. The combined process is more effective as compared to a single process. Yu et al. [35] combined physical, chemical, and biological pretreatment process to pretreat rice husk. Results indicate that the combination of chemical (2% H_2SO_4) and biological (*P. ostreatus*) pretreatments leads to a higher lignin degradation than single-step pretreatments. The combined pretreatment of napier grass carried out using 2% NaOH along with cellulase enzyme was found to be more effective as compared with single alkaline pretreatment, in which a 3.97 time higher methane production (MP) was obtained [36].

3. Potential bio-hydrogen and methane production from lignocellulosic biomass

Alternative fuels are recently in high demand owing to concerns about depletion of fossil fuels and harmful gases emission problem which results in climate change and environmental deterioration [37]. Biofuels (fuel alcohol, biodiesel/bio-jet, and biogas) can be a suitable alternative to fossil fuels as they are derived from renewable feedstocks, biodegradable, and combusted based on carbon dioxide cycle [38]. Biofuels can be used for the energy generation by combustion or other technologies. They have been used in transportation and power generation sectors, in which the share of biofuel in transport fuel demand has been increasing and reached 3% in 2017 [39]. Biogas (hydrogen and methane) is a highly promising biofuel because it can be produced from a variety of organic feedstocks, including waste biomass which can attribute to the waste reduction simultaneously with energy production [40].

3.1 Hydrogen

Hydrogen is a noncarbonaceous fuel and energy carrier possessing higher net calorific value compared to other fuels [41]. It can be directly converted into energy in fuel cell or mixed with natural gas for use in internal combustion and jet engines, as well as the gas power turbines. Combustion of hydrogen yields only water; thus it is considered as a clean energy source. The limitation in using hydrogen is its explosivity when mixed with oxygen, leading to difficulty in its storage and distribution [42]. Production of hydrogen from lignocellulosic biomass can be achieved by gasification and microbial fermentation technologies. Gasification is very energy-intensive and releases large amount of carbon, sulfur, and nitrogen oxides to the atmosphere [43]. Therefore, attention had been paid to the microbial fermentation process as it is more environmentally friendly. Bio-hydrogen is a term used to call hydrogen produced via microbial fermentation. Dark- and photo-fermentation are typically applied for bio-hydrogen production. Dark fermentation of organic carbon substrates is carried out by obligate or facultative anaerobic bacteria yielding bio-hydrogen and other side products, such as volatile fatty acids (VFAs) and alcohols. Photo-fermentation requires energy from light to aid the decomposition of organic substrates by photosynthetic bacteria, mostly purple non-sulfur bacteria (PNSB) [44]. The dark fermentative bacteria are capable of utilizing various substrates with high rate of hydrogen production. A drawback of dark fermentation is its low yield due to the large quantity of side products formed. The substrates for photo-fermentative bacteria are limited to simple sugars and organic acids, and the hydrogen production rate by photo-fermentation is usually lower than dark fermentation [44]. However, with the high substrate conversion efficiency and high hydrogen yield (HY), the photo-fermentation is also considered a promising technology for bio-hydrogen production [45]. In addition, recent research reported the sequential dark-photo-fermentation as an efficient bio-hydrogen production process. The VFAs from dark fermentation are further utilized for hydrogen production in photo-fermentation, thus the HY and substrate conversion efficiency can be improved via sequential dark-photo-fermentation [45–47]. Typical reactions for dark fermentation with acetic acid formation and photo-fermentation can be stated as follows:

$$\text{Dark fermentation}: C_6H_{12}O_6 \rightarrow 2CH_3COOH + 2CO_2 + 4H_2 \quad (1)$$

$$\text{Photo-fermentation}: 2CH_3COOH + 4H_2O + \text{Light} \rightarrow 4CO_2 + 8H_2 \quad (2)$$

Despite the continuing research at the laboratory scale, the biological hydrogen production from lignocellulosic biomass at pilot and industrial scales is still limited. Various kinds of lignocellulosic feedstock have been investigated for bio-hydrogen production by different microorganisms. Typically, the feedstocks are pretreated prior to fermentation in order to enhance hydrogen production efficiency. Pretreatment of the biomass can be conducted by physical (such as size reduction), physicochemical (such as steam, ammonia fiber, and carbon dioxide explosion, hot water, and microwave pretreatment), chemical (such as alkaline, diluted acid, and hydrogen peroxide pretreatment), and biological (such as enzymatic pretreatment) methods. The yield of hydrogen from lignocellulosic feedstocks is diverse depending on the types of substrates, pretreatment methods and microorganisms used. Under mesophilic condition, dark fermentation of untreated water hyacinth by mixed culture of *Enterobacter* sp. and *Clostridium* sp. resulted in 119.6 mL-H_2/g-VS [48]. Enzymatic hydrolysates of agave bagasse yielded 1.53–3.40 mol-H_2/mol-substrate by anaerobic mixed cultures [49, 50]. Higher hydrogen production from acid hydrolysate of sugarcane bagasse (6980 mL-H_2/L-substrate) was observed with mixed cultures compared to the pure culture of *Enterobacter aerogenes* (1000 mL-H_2/L-substrate) [51, 52].

The pretreated lignocellulosic biomass (in solid form) can also be directly fermented to hydrogen. Alkaline-pretreated sugarcane bagasse fermentation *by C. beijerinckii* yielded 0.733 mmol-H_2/g-substrate [53]. The HY of 51.9 mL-H_2/L-substrate was obtained by fermenting corn stover obtained after steam explosion using mixed cultures of *C. celluloblyticum* and *Citrobacter amalonaticus* [54]. The pretreated solid biomass could also be used as feedstocks for hydrogen production via simultaneous saccharification and fermentation (SSF) process. The cellulolytic enzymes mostly perform well under thermophilic condition (50–60°C). However, hydrogen production by SSF under mesophilic condition had been investigated by some researchers based on the optimal temperature for growth and activity of hydrogen producers. Hydrogen yield of 72 mL-H_2/g-substrate was obtained from acetic acid steam-exploded corn straw by SSF with *Ethanoligenens harbinense* at 37°C [55]. A lower yield of 68 mL-H_2/g-substrate was obtained from steam-exploded corn straw by SSF with *C. bytyricum* AS1 at 35°C [56].

Fermentation under thermophilic condition (50–65°C) was reported to improve dark fermentative hydrogen production via enhancing substrate degradation rate. Various thermophilic hydrogen producers, such as *Thermoanaerobacterium thermosaccharolyticum* [57, 58], *C. thermosaccharolyticum*, and *C. thermocellum* [59, 60], as well as thermophilic mixed cultures [61], were applied for hydrogen production from lignocellulosic feedstocks and their hydrolysates. HY of 1947 mL-H_2/L-substrate from microwave-assisted acid hydrolysate of oil palm trunk (OPT) was achieved using *T. thermosaccharolyticum* KKU19 [62], while the enzymatic hydrolysate of lime-pretreated OPT yielded 2179 mL-H_2/L-substrate using the same strain [58]. Corn stover hydrolysate obtained by diluted sulfuric acid pretreatment was fermented by *T. thermosaccharolyticum* W16 with a yield of 2.24 mol-H_2/mol-sugar [63]. When the enzymatic hydrolysate of NaOH-pretreated corn stover was used, the strain W16 produced 108.5 mmol-H_2/L-substrate [64]. Solid residues of sweet sorghum stalk after hydrogen fermentation was subjected to diluted sulfuric acid hydrolysis. The resulting acid-treated slurry was further fermented by *C. thermosaccharolyticum* DSM572 and yielded 2.5 mmol-H_2/g-substrate [65]. Activated sludge and anaerobic granular sludge produced 627 and 822 mL-H_2/L-substrate from diluted sulfuric acid hydrolysate of corn stover under thermophilic condition, which were 2.3 and 3.7 times higher than those obtained under mesophilic condition [61]. Sweet sorghum stalks were used as substrate for hydrogen production by mixed cultures of *C. thermocellum* DSM7072 and *C. thermosaccharolyticum* DSM572. The HY of 5.1 mmol-H_2/g-substrate was observed [59]. Fermentation of hydrogen by thermophilic microorganisms could overcome the technical challenge of SSF regarding difference between optimal temperatures for enzymatic saccharification and fermentation. SSF of lime-pretreated OPT by *T. thermosaccharolyticum* KKU19 achieved a maximum yield of 60.22 mL-H_2/g-pretreated OPT [66]. Fungal-pretreated cornstalk yielded 89.3 mL-H_2/g-substrate by SSF process with *T. thermosaccharolyticum* W16 [67].

Co-digestion with nitrogen-rich organic biomass was reported to enhance hydrogen production from lignocellulosic feedstocks. The OPT hydrolysate co-digested with slaughterhouse wastewater by *T. thermosaccharolyticum* KKU19 gave 2604 mL-H_2/L-substrate [68]. Co-digestion of napier grass and its silage with cow dung with the bioaugmentation of *C. butyricum* TISTR 1032 yielded 6.98 and 27.71 mL-H_2/g-volatile solid (VS) [69]. Wheat straw and cheese whey were co-digested by anaerobic granular sludge, and the hydrogen production of 4554, 3685, and 4132 mL-H_2/L-substrate were observed in 0.11-L serological bottle, 1-L bioreactor, and 4-L bioreactor, respectively [70].

Photo-fermentative hydrogen production mostly uses simple sugars (such as glucose) or organic acids (such as acetic and butyric acids) as substrates. Lignocellulosic hydrolysates with sugar monomers were investigated for hydrogen production. Photo-fermentation of enzymatic hydrolysate of ammonia pretreated wheat straw by *Rhodobacter capsulatus*-PK gave 712 mL-H_2/L-substrate [71]. Corn stalk pith was hydrolyzed by cellulase enzyme. The resulting hydrolysate was fermented by photosynthetic consortium comprising *R. capsulatus*, *R. sphaeroides*, *Rhodopseudomonas capsulata*, *Rhodopseudomonas palustris*, and *Rhodospirillum rubrum*, in which a HY of 2.6 mol-H_2/mol-sugar consumed was achieved [72].

Sequential dark-photo-fermentation was applied to increase HY from lignocellulosic biomass. The organic acids obtained from dark fermentation of lignocellulosic biomass are used as substrate for photo-fermentation. The yield of hydrogen from water hyacinth was enhanced from 76.7 to 596.1 mL-H_2/g-total volatile solid (TVS) by combining dark fermentation (using mixed hydrogen-producing bacteria) with photo-fermentation (using *R. palustris*) [73]. Dark fermentation of pretreated corn stalk by mixed culture from cow dung yield 192.9 mL-H_2/g-TVS. The yield was increased to 401.5 m mL-H_2/g-TVS by *combining* with photo-fermentation using *R. sphaeroides* HY01 [74]. Yang et al. [75] reported a HY from pretreated corncob by dark fermentation with mixed cultures from dairy manure of 120.2 mL-H_2/g-corncob. Photo-fermentation of the effluent from this process gave 713.6 mL-H_2/g-COD. The authors also stated that reducing sugars and oligosaccharides in corncob hydrolysate contributed to the hydrogen produced by dark fermentation, while acetic acid, butyric acid, and alcohols in the dark fermentation effluent contributed to the hydrogen produced by photo-fermentation [75]. A pilot scale test of sequential dark-photo-fermentation from corn stover was investigated. Sewage sludge and photo-hydrogen-producing consortia HAU-M1 were used as inoculum for dark and photo-fermentation, respectively. The overall volumetric hydrogen production rate (HPR) was 7.8 and 4.7 m^3/ m^3d from dark and photo-fermentation, respectively [47].

3.2 Methane

Methane is a fuel gas mainly produced from anaerobic digestion process. Organic substrates are decomposed by diverse microbial communities through a series of metabolic stages during anaerobic digestion, resulting in gaseous products called biogas and inorganic molecules remaining in digestate. Biogas mainly comprises methane (50–75%), carbon dioxide (25–40%), nitrogen (<5%), hydrogen (<1%), oxygen (<1%), and hydrogen sulfide (50–5000 ppm) [76]. Biogas is suitable for use in internal combustion engines and gas turbine generators. Methane has higher octane rating than gasoline, and its combustion produces less CO_2 as compared to fossil fuels [77].

Methane production by anaerobic digestion process involved multiple steps performed by several groups of microorganisms. Typically, anaerobic digestion is divided into four steps that are hydrolysis, acidogenesis, acetogenesis, and methanogenesis. In hydrolysis step, complex organic matters (such as cellulose and protein) are converted into simpler and soluble molecules (such as sugars and amino acids) by hydrolase enzymes excreted by facultative and strictly anaerobic microorganisms called fermentative bacteria. The soluble molecules produced by the hydrolysis steps are then utilized by acidogenic bacteria to produce short-chain organic acids (such as acetic, butyric, and propionic acids) along with hydrogen, carbon dioxide, and alcohols in the acidogenesis step. These products are further consumed in the acetogenesis step to produce acetic acid by acetogenic bacteria. In the last step, acetic acid, hydrogen with carbon dioxide, formic acids, and alcohols were utilized by methanogenic bacteria to produce methane under obligate anaerobic condition [76]. The optimal condition for methanogenic bacteria and other groups of bacteria are different. Some researchers, therefore, introduced two-stage hydrogen and methane production carried out by separating the fermentation into two phases of acidogenesis and methanogenesis, which can promote the methane fermentation rate and increase energy yield from feedstocks [27, 58].

Methane production from various lignocellulosic biomasses has been investigated by different research groups. Due to their complex structures which limit the bioavailability, hydrolysis was reported as the rate-limiting step for methane production from lignocellulosic feedstocks [78]. In order to increase methane production rate (MPR) and improve methane production efficiency, different pretreatment methods such as size reduction, thermal, hydrothermal, alkaline, dilute acid, thermal alkaline/dilute acid, and fungal pretreatments were applied [79–82]. Alkaline pretreatment and combination of alkaline with other pretreatment methods are usually employed, while the thermal pretreatment is reported as the suitable method resulting in greater than 50% increased methane yield compared to un-pretreated feedstocks [83].

Theoretically, the yield of methane, at standard temperature and pressure, from cellulose and hemicellulose are 415 and 424 mL-CH_4/g with 50% methane content in the biogas [76]. Since the compositions of lignocellulosic biomass are diverse, the yield of methane varies depending on the type lignocellulosic feedstocks used.

Herbaceous biomasses are common lignocellulosic feedstocks for methane production. Corn stover yielded 320–335 mL-CH_4/g-VS [79, 84]. Co-digestion of corn stover with goose manure increased the methane yield (MY) to 393 mL-CH_4/g-VS [85]. The straws of wheat, rice, and corn gave 240–329 mL-CH_4/g-VS [86–91]. Relatively low values HY were observed from biomass of grasses (142–301 mL-CH_4/g-VS) [36, 83, 92, 93] and woody biomass (136–205 mL-CH_4/g-TS) [79, 94], while bagasse feedstocks yielded relatively high values of 330–420 mL-CH_4/g-VS [95, 96].

Two-stage hydrogen and methane production was reported as a successful process to produce hydrogen together with methane and enhance energy recovery from lignocellulosic biomass. Energy yield from OPT hydrolysate increased from 0.8 to 10.6 kJ/g-COD by applying two-stage thermophilic hydrogen and mesophilic methane production in comparison to one-stage thermophilic hydrogen production [57]. The HY of 53.8 mL-H_2/g-VS together with HY of 133.9 mL-CH_4/g-VS was achieved by two-stage fermentation of maize silage [97]. Sequential hydrogen and methane fermentation of sugarcane bagasse hydrolysate obtained by steam explosion yielded a total energy of 304.11 kJ/L-substrate [78]. The gaseous (hydrogen and methane) recovery from mixed sugarcane bagasse hydrolysate and water hyacinth was maximized by continuous two-stage hydrogen and methane production at a hydraulic retention time of 8 h and 10 days, respectively, providing energy yield of 8.97 KJ/g-COD [98]. Continuous two-stage hydrogen and methane production from agave bagasse enzymatic hydrolysate was optimized at an organic loading rate of 44 g-COD/L-d (for hydrogen) and 20 g-COD/L-d (for methane), in which 9.22 kJ/g-bagasse was recovered [99].

4. Factors affecting dark fermentative hydrogen production

4.1 Types of inoculum

Pure and mixed cultures are two types of inoculum used to produce hydrogen by dark fermentation. *Clostridium* sp. and *Enterobacter* sp. are the pure culture widely used to produce hydrogen. Pure cultures give the high HPR and HY [100]. The major disadvantage of using pure culture is the sterile conditions which are required during the start-up and operations resulting in high operation costs from an energy use. This problem can be mitigated by using mixed cultures. Using mixed cultures as an inoculum in bio-hydrogen fermentation process is more practical than those using pure culture because it is simpler to operate, the process is easier to be controlled [101], and its feasibility to use complex organic wastes [100]. Inoculum sources for mixed cultures are animal dung, anaerobic sludge, municipal solid waste, soil, and compost [102]. The presence of hydrogen consumers such as methanogens and homoacetogens is the drawbacks of using mixed cultures. In order to inhibit these hydrogen consumers while harvesting the hydrogen producers, the pretreatment methods including heat treatment; acid treatment; alkali treatment; sonication; aeration; freezing and thawing; addition of specific chemical compounds, e.g., 2-bromoethanesulfonic acid; and addition of long-chain fatty acids are needed [103, 104].

4.2 Feedstocks

Various kinds of feedstock have been used to produce hydrogen by dark fermentation. They can be classified into three generations. First-generation feedstocks are food crops such as sugarcane, sugar beet, corn, and cassava which can be easily digested by fermentative microorganisms. However, there is the concern on food competition and arable land when food crops are used to produce biofuels [105]. Thus, lignocellulosic biomass is developed as the second-generation feedstocks. Due to its compositions, lignocellulosic biomass is difficult to be digested by microorganisms. Therefore, the pretreatment and hydrolysis of the lignocellulosic biomass are needed in order to obtain its underlying monosugars prior the fermentation. Recently, the third-generation feedstock, i.e., microalgae has received high attention to produce hydrogen. Microalgae have rapid growth rate with a high capturing ability for CO_2 and other greenhouse gases. They can be cultivated without soil and have a very short harvesting cycle (1–10 days) [106, 107]. Microalgae biomass consists of high carbohydrates (cellulose and starch) and lipid contents that can be converted to hydrogen by hydrogen producers. HY, HPR, and the overall economy of the process [102] are affected by the differences in carbohydrate content, bioavailability, and biodegradation rate of the first-, second-, and third-generation feedstocks. In addition, the concentrations of feedstocks must be considered because a feedstock or product inhibition can occur in the fermentation process [108].

4.3 Nitrogen and phosphate

Nitrogen is required for growth of hydrogen-producing bacteria. Nitrogen source for fermentative hydrogen production is classified into inorganic and organic. Examples of inorganic nitrogen are ammonia nitrogen [109], ammonium bicarbonate [110], and ammonium chloride [111, 112]. Ammonia nitrogen is the most widely used inorganic nitrogen with its optimal concentration in the range 0.1 to 7.0 g/L [113, 114]. Peptone, yeast extract, and corn steep liquor are the examples of organic nitrogen. Ferchichi et al. [115] and Ueno et al. [116] reported that a higher HY was obtained when organic nitrogen is supplied to the fermentation medium.

In fermentative hydrogen production, phosphate is needed due to its nutritious value as well as buffering capacity. An increase in phosphate concentration results in increase of the capability of the bacteria to produce hydrogen. However, too high concentrations of phosphate could cause the substrate inhibition [113, 117]. The optimum C/N and C/P ratios are 74:200 and 599:1000, respectively [118, 119].

4.4 Temperature

Temperature affects the maximum specific growth rate, substrate utilization rate, hydrolysis of the substrate, mass transfer rate, hydrogen partial pressure, hydrogenase activity, and the metabolic pathway of the bacteria resulting in a shift of by-product compositions [101, 120, 121]. Fermentative hydrogen production can be operated under a wide range of temperature, i.e., mesophilic (25–40°C), thermophilic (40–65°C), or hyperthermophilic (>80°C) ranges [122]. Thermophilic condition gave a higher hydrogen production than the mesophilic condition. Sotelo-Navarro et al. [123] reported that the bio-hydrogen production from disposable diapers at 55°C was greater at 35°C. This could be due to the increased pace of microbial metabolism in the thermophilic condition. The optimal temperature for fermentative hydrogen production varies depending on the inoculum and substrate types.

4.5 pH

pH affects the activity of hydrogenase as well as the metabolism pathway of the microorganisms [109]. Low pH inhibited hydrogenase activity [124, 125] resulting in longer lag time [126] and the inhibition of dark fermentation process. This can be attributed to the protonation of undissociated acids in medium which can penetrate the microbial cell membrane and inhibit the growth and activities of microorganism [127]. Acidic pH of 4.5–6.0 favors the acetic and butyric acid production pathway. High initial pH leads to the production of ethanol and propionate rather than hydrogen production [128]. The propionate production pathways consume reducing powers that are potentially used for hydrogen synthesis [108].

4.6 Metal ion

Fe, Ni, and Mg are required for bio-hydrogen production process. These metals are cofactors for enzymes facilitating transport processes in the microorganisms [122, 129, 130]. Fe^{2+} is an important element to form hydrogenase and other enzymes. Fe-S affects protein functions by acting as an electron carrier and involving in oxidation of pyruvate to acetyl-CoA, CO_2, and H_2 [122]. Additionally, Fe^{2+} induces metabolic alteration and is involved in Fe-S and non-Fe-S protein operation in hydrogenase [122, 131]. Nickel is a fundamental component of [NiFe]-hydrogenase. It has the influences on the activity of [NiFe]-hydrogenase. High concentration of nickle inhibits the activity of [NiFe]-hydrogenase, promoting fermentative hydrogen production [122, 132]. Mg^{2+} is an element that is found abundantly in microbial cells. It stabilizes ribosomes, cell membranes, and nucleic acids and plays a crucial role as an activator of many kinases and synthetases [133]. Cu, Cr, and Zn also have influences on hydrogen fermentation process [122]. The relative toxicity of these heavy metals are Zn (most toxic) > Cu > Cr (least toxic).

4.7 Hydraulic retention time

Hydraulic retention time (HRT) is defined as the time that fermentation broth remains in a reactor. It is related to the working volume of the reactor and the influent flow rate. HRT affects a continuous hydrogen production. Hydrogen-producing bacteria are fast-growing bacteria, so they prefer short HRT, while the methanogens are slow-growing microorganisms, so they prefer long HRT [134]. Therefore, HRT can be used as controlling parameters to suppress the community of methanogens [102]. Jung et al. [134] reported that the HRT for treating liquid-type substrate is shorter than that of solid-type substrate because the times to hydrolyze substrate containing high solid are much longer.

4.8 Hydrogen partial pressure

Hydrogen partial pressure affects hydrogenase activity because it is involved in reversibly oxidizing and reducing ferredoxin [102]. High accumulation of hydrogen partial pressure in the fermentation broth decreases the hydrogen production because the reaction tends to be reducing ferredoxin rather than oxidizing ferredoxin [135]. Hydrogen partial pressure can be reduced by biogas sparging [136], agitation, and reduction of headspace pressure using vacuum pump or enlarging the headspace volume.

5. Factors affecting photo-fermentative hydrogen production

5.1 Carbon sources

Various kinds of substrates can be used as carbon source by PNSB. Short-chain organic acids such as malic, lactic, succinic, acetic, propionic, and butyric acids

[137–141] are the most generally used substrates for photo-hydrogen production. VFAs in the hydrogenic effluent can also be used to produce hydrogen by PNSB [142–145]. Additionally, other carbohydrate substrates [37, 146, 147] and organic acids from industrial wastewaters can be utilized as carbon source by PNSB [148–151]. Carbon affects the metabolism of cell growth and photo-hydrogen fermentation system [152, 153]. Cell formation utilizes large fraction of carbon, while hydrogen production utilizes a smaller fraction. The efficiency of photo-hydrogen production is different according to the types of carbon substrates. This is due to the variations in the electron transfer capabilities in the different metabolic pathways of photosynthetic microbes [154]. Substrate concentration can also affect the photo-hydrogen production. The optimum concentrations of VFAs for photo-hydrogen production were reported in the range of 1800–2500 mg/L [155, 156]. The maximum theoretical HY from different carbon substrates are as follows:

$$\text{Lactate}: C_3H_6O_3 + 3H_2O \rightarrow 6H_2 + 3CO_2 \tag{3}$$

$$\text{Malate}: C_4H_6O_5 + 3H_2O \rightarrow 6H_2 + 4CO_2 \tag{4}$$

$$\text{Butyrate}: C_4H_8O_2 + 6H_2O \rightarrow 10H_2 + 4CO_2 \tag{5}$$

$$\text{Acetate}: C_2H_4O_2 + 2H_2O \rightarrow 4H_2 + 2CO_2 \tag{6}$$

$$\text{Propionate}: C_3H_6O_2 + 4H_2O \rightarrow 7H_2 + 3CO_2 \tag{7}$$

$$\text{Formate}: CH_2O_2 \rightarrow H_2 + CO_2 \tag{8}$$

5.2 Nitrogen sources

Nitrogen is an essential nutrient for cell synthesis and hydrogen production. The activity of nitrogenase, an enzyme involved in the hydrogen production by photosynthetic bacteria, is greatly affected by nitrogen. Glutamate is a preferred nitrogen source for PNSB. It was rapidly consumed and could also improve hydrogen production of photo-hydrogen-producing bacteria [157–159]. Ammonia has an adverse effect on hydrogen production. High concentration of ammonium ions powerfully inhibited the synthesis and activity of nitrogenase. However, a low ammonium concentration less than a non-inhibitory level can support the growth of cells and is able to enhance the photo-hydrogen production.

5.3 pH

pH affects the ionic concentration in the medium. These ionic forms influence the active site of nitrogenase and affect the biochemical characteristic in microbial cells during metabolism process [154, 160]. Optimal pH for photo-hydrogen production of PNSB was 7.0 [140, 161–164].

5.4 Temperature

An increase in the environmental temperature until the optimal temperature can improve the activities of the nitrogenase and proteins associated with the cell growth or hydrogen production. An imbalance of incubation temperature on cells growth inhibits the physiological activity, intracellular enzyme activity, and metabolism of cells. Unstable temperature may cause bacteria to spend their energy for adaptation to low/high temperatures in order to be able to survive [165] which results in a reduction in the hydrogen production, HPR, HY, and substrate conversion efficiency [139, 154, 166].

5.5 Light energy

Light energy is a necessary resource for the reaction, electron transport, ATP synthesis, and hydrogen production [165, 167]. Light intensity influences the HPR and cell synthesis [160, 163, 168]. At the optimal light intensity, large amounts of ATP and reductive power are sufficient for nitrogenase activity to produce hydrogen and generate the cells. However, a further increase in light intensity greater than the saturation condition became an inhibitory for hydrogen production by PNSB. Photo inhibition occurs when the photosynthetic system supplies excess ATP and Fd_{red} in comparison to the capacity of nitrogenase enzyme [169]. Consequently, the cell is damaged by the bleaching bacteriochlorophyll pigment during the extra-light cultivation [170].

Halogen [141, 152, 171], tungsten [155, 161], fluorescent [172], infrared [172], and light-emitting diode (LED) lamps [173, 174] have been used as the light source for photo-hydrogen fermentation. Among these lamps, LED has the high operational stability and can improve the performance of photo-hydrogen fermentation [154]. Other advantages of LED include specific wavelengths (770–920 nm), lower electricity consumption, lower heat generation, and longer life expectancy [154, 174].

5.6 Iron concentration

Iron is the major cofactor at the active site of FeMo-nitrogenase [157, 175]. There are 24 atoms of Fe as the composition in each molecule of nitrogenase [176]. It is also an essential component in ferredoxin and cytochrome b-c complex, which are electron carriers of the photosynthetic electron transport system. Ferredoxin also contains Fe_4S_4 in a cluster of nitrogenase [177]. Photo-hydrogen production is functioned by nitrogenase, which receives electron carriers from ferredoxin and reduces protons to molecular hydrogen. The optimal Fe^{2+} concentration for photo-hydrogen fermentation are in the range of 1.68–35 mg/L [164, 177–179]. Concentration of iron greater than the requirement of regular physiological metabolisms can disrupt the cell surface of microorganisms. As a consequence, the production of hydrogen is reduced [177].

5.7 Vitamin solution

Vitamins are essential for carbohydrate, protein, lipid, and cell metabolism [180, 181]. Vitamin B1 (thiamine) is a precursor of thiamine pyrophosphate (TPP), a coenzyme of the pyruvate dehydrogenase complex, essential for catabolism of carbohydrates, organic acids, and amino acids. This is important in the conversion of pyruvic acid and provides acetyl-CoA in the TCA cycle which supports cell synthesis. Biotin is a part of an enzymatic carboxylation and is a cofactor for carbon dioxide fixing enzymes such as pyruvate carboxylase. Oxaloacetate is supplied by pyruvate carboxylase. This is important in the citric acid cycle and in the production of biochemical energy. Vitamin B6 (pyridoxamine) is necessary for the metabolism of amino acid and in glycogen hydrolysis [181–183]. Nicotinic acid is a precursor of NAD^+/NADP, which are electron carrier and play an important role in electron transfer during the photo-fermentation process [180].

5.8 Inoculum concentration

The ratio of initial cell concentration (X_0) to initial substrate concentration (S_0) affects the initial energy level of microorganisms. This energy is necessary to support the cell synthesis and metabolism process [140]. At a high S_0/X_0 ratio, i.e., low seed inoculation, microorganisms require more adaptation to utilize the high substrate concentration leading to a delay of the lag period for photo-hydrogen fermentation [140, 162, 163]. A further increase in cell concentration to greater than the optimal level resulted in a decreased hydrogen production [140, 162, 163]. A low S_0/X_0 ratio, i.e., high seed inoculation, resulted in an insufficiency of the substrate to supply the growth of cells [140]. In addition, excess biomass prevents penetration of light into the cultivation system due to a self-shading effect. This leads to a decrease in light intensity that causes a reduction of ATP creation resulting in the decrease of hydrogen production by photosynthetic bacteria. Moreover, extracellular concentrations might promote the formation of bacterial flocs or biofilm creation which can limit substrate distribution inside the bioreactor system [140, 157, 167].

6. Factors affecting methane production

6.1 pH

pH influences the growth of microorganisms in various stages of the anaerobic digestion (AD) process [184, 185]. Optimum pH for methanogens to produce methane ranges from 7.0 to 7.2 [186]. pH outside the range of 6.0–8.5 is toxic to methanogens. pH values below 6.6 starts to adversely affect the activities of the methanogens, and the values below 6.2 are significantly toxic to the methanogens. During the acidogenesis stage of AD process, the pH in an anaerobic digester decreases to below 6.0 due to VFAs accumulation and carbon dioxide production. After this, the pH rises to 7.0–8.0 or above. Yu and Fang [186] and Kim et al. [187] found that the optimal hydrolysis and acidogenesis stage were achieved at pH 5.5–6.5, and the acidogenic bacteria continue to produce the acids until the pH drops to 4.5–5.0 [186, 188, 189]. As a consequence, the activity of methanogens is inhibited. Thus, it is recommended that the hydrolysis, acidification, acetogenesis/methanogenesis stage in AD process should be carried out separately [190].

6.2 Temperature

Most of the methanogens are mesophile which are active in the temperature ranges of 30–35°C, while only a few are thermophile which are active in the temperature ranges of 50–60°C [186]. Deublein and Steinhauser [190] reported that the methanogenic activity is inhibited at the temperatures between 40 and 50°C especially at the values near 42°C. This is believed to be a transition temperature from mesophilic to thermophilic temperature [191].

6.3 Hydraulic retention time

HRT affects the rate and extent of methane production. A long HRT results in higher total VS mass reduction, which in turn leads to higher cumulative biogas production as well as to allow the microorganisms to acclimate to toxic compounds [191]. Methanogens have a long generation time. Thus, the HRT is usually set at 10–15 days to avoid the washout from the reactor [186]. The length of HRT can vary depending on bacterial stains, operation condition, and so on.

6.4 Substrate composition

Biogas yield and the compositions of biogas are greatly influenced by the composition of feedstocks. AD of carbohydrates, fats, and protein yield 886, 1535, and 587 L biogas/kg-VS with methane content of approximately 50, 70, and 84%, respectively, [184]. Substrate to inoculum ratio (S/I), as well as biodegradability of the substrate, is another important factor affecting batch AD processes, especially at high solid content [192]. Too high S/I ratio may be toxic, while too low S/I ratio may prevent induction of the enzyme necessary for biodegradation [193]. Too high concentration of feedstock can cause inhibition or failure of AD [194] due to substrate inhibition. High S/I ratio can lead to overloads due to VFAs accumulation [192, 195] and long lag phase. Thus, a low S/I ratio is preferred in order to attain shorter lag phase [192, 196]. Owen et al. [197] proposed a standard S/I ratio to be approximately 1 $g\text{-}VS_{substrate}/g\text{-}VS_{inoculum}$.

6.5 Organic loading rate

Organic loading rate (OLR) is defined as the amount of VS or COD components fed per day per unit digester volume. Higher OLR can reduce the digester's size and the capital cost as a consequence. However, enough time (HRT) should be provided to the microorganisms for breaking down the organic material and converting it into gas [198]. An increase in OLR can result in higher hydrogen production efficiency [199]. However, a further increase in OLR beyond a certain level will result in substrate inhibition, leading to a lower MY [200]. Too high OLR can shift the metabolic to solventogenic phase [201]. Hobson and Bousfield [201] and Chandra et al. [185] reported that a total solid content of 8.0–10.0% is desirable for optimum MY.

6.6 Alkalinity

Buffer capacity, or alkalinity in AD process, is the equilibrium of carbon dioxide and bicarbonate ions that provides resistance to significant and rapid changes in pH. Alkalinity is proportional to the concentration of bicarbonate. The imbalance of digester can be more reliably measured by a buffer capacity than a direct measurement of pH. This is because an accumulation of short-chain fatty acids will reduce the buffering capacity significantly before the pH decreases. A low buffer capacity can be improved by reducing OLR. This is because too high OLR inhibited the microorganisms due to too high fatty acid concentration. Reducing OLR therefore reduces the availability of these fatty acids. Then, the alkalinity of the system can be improved. Guwy et al. [202] and Ward et al. [189] indicated that a quicker way is to add strong bases or carbonate salts to remove carbon dioxide from the gas space and convert it to bicarbonate. Alternatively, bicarbonate (HCO_3^-) can be added directly.

6.7 Carbon to nitrogen (C/N) ratio

C/N ratio is a ratio of the mass of carbon to the mass of nitrogen in a substance. In AD process, a C/N ratio ranging from 20 to 30 is considered optimum [184, 185], and the value of at least 25:1 is suggested for optimal gas production [191]. If the C/N ratio is too high, methanogens will rapidly consume the nitrogen to meet their protein requirements and will no longer react with the rest of carbon in the material. As a result, gas production will be low. On the other hand, if the C/N ratio is too low, nitrogen will be liberated and accumulated in the system in the form of ammonium ion (NH_4^+). This can possibly increase the pH of the digestate to a level that is toxic to methanogens (pH 8.5) [186, 203].

7. Processes for bio-hydrogen and methane production from lignocellulosic materials

The abundance of lignocellulosic biomass makes it a viable feedstock for hydrogen (H_2) and methane (CH_4) production. Cellulose in lignocellulosic biomass can be saccharified to glucose then fermented to hydrogen and methane. In this section, summarized details on fermentative conversion process for hydrogen, i.e., dark fermentation and photo-fermentation, methane production, and AD are given.

7.1 Processes for fermentative hydrogen production

The methods that are investigated widely for fermentative hydrogen production are dark fermentation, photo-fermentation, and a coupling system comprising dark fermentation and photo-fermentation [204]. Dark fermentation is an acidogenic fermentation process conducted under anaerobic conditions in the absence of light. Dark fermentation, as compared to photo-fermentation, is regarded as a more promising method [42], owing to its ability to utilize a wide range of biomass, its high hydrogen production rate, and its independence of lighting conditions [109]. Microorganisms used in dark fermentation are strictly anaerobic bacteria, particularly those in the genus *Clostridium*, and facultative anaerobic bacteria, e.g., *Enterobacter* spp. [205]. Mixed cultures, for example, sludge compost and sewage sludge, are also used [204]. In theory, the maximum HY obtained under dark fermentation is 4 mol-H_2/mol-glucose when acetic acid is produced as a co-product (Eq. (9)). This is roughly equivalent to one third of energy recovery from the biomass [204]. The HY of 2 mol-H_2/mol-glucose can also be obtained when butyric acid is produced as the co-product (Eq. (10)). However, when mixed culture is used, mixed acids are often produced, leading to a lower HY of 2.5 mol-H_2/mol-glucose (Eq. (11)).

$$C_6H_{12}O_6 + 6H_2O \rightarrow 2CO_2 + 2CH_3COOH + 4H_2 \tag{9}$$

$$C_6H_{12}O_6 + 6H_2O \rightarrow 2CO_2 + CH_3CH_2CH_2COOH + 2H_2 \tag{10}$$

$$4C_6H_{12}O_6 + 6H_2O \rightarrow 3CH_3CH_2CH_2COOH + 2CH_3COOH + 8CO_2 + 10\ H_2 \tag{11}$$

Photo-fermentation is another process being investigated widely for hydrogen production from biomass. Unlike dark fermentation, photo-fermentation is a process that requires light to drive the conversion of organic substrates into hydrogen. Purple non-sulfur bacteria are a group of microorganisms responsible for hydrogen production under photo-fermentation. Examples of PNSB include *Rhodobacter* spp., *Rhodopseudomonas* spp., and *Rhodospirillum* sp. Photo-fermentation is a process known for its high substrate conversion efficiencies [206]. In theory, photo-fermentation can completely convert organic compound into hydrogen, i.e., 12 moles of hydrogen can be obtained from a mole of glucose (Eq. (12)), which is much higher than that obtained through dark fermentation (4 mol-H_2/mol-glucose). However, when VFAs are used as the substrate, lower HYs in a range 1–10 mol-H_2/mol-VFA are obtained (Eqs. (13)–(17)). In photo-fermentation, it was reported that PNSB showed an affinity toward VFAs, with malate and lactate being the most preferable substrate. Nevertheless, a good yield is also reported using acetate as the substrate [206].

$$C_6H_{12}O_6 + 6H_2O + Light \rightarrow 6CO_2 + 12H_2 \quad (12)$$

$$HCOOH + Light \rightarrow CO_2 + H_2 \quad (13)$$

$$CH_3COOH + 2H_2O + Light \rightarrow 2CO_2 + 4H_2 \quad (14)$$

$$CH_3CH(OH)COOH + 3H_2O + Light \rightarrow 3CO_2 + 6H_2 \quad (15)$$

$$HO_2CCH(OH)CH_2COOH + 3H_2O + Light \rightarrow 4CO_2 + 6H_2 \quad (16)$$

$$CH_3CH_2CH_2COOH + 6H_2O + Light \rightarrow 4CO_2 + 10H_2 \quad (17)$$

Due to the ability of photo-fermentation to utilize VFAs as the substrate for hydrogen production, in recent years, much attention has been paid on improvement of hydrogen production from biomass using coupling systems comprising dark fermentation and photo-fermentation. Anaerobic bacteria and PNSB can be co-cultivated in a single bioreactor, so that VFAs produced as the co-products during dark fermentation are instantly converted into hydrogen by photo-fermentation. Several co-cultivation of anaerobic bacteria, either pure or mixed culture, and PNSB have been reported in literatures with better HYs compared with the use of single-strain cultivation, for example, *C. butyricum* and *Rhodobacter* sp. M-19 [207], *C. butyricum* and *R. sphaeroides* [208], and *Lactobacillus delbrueckii* and *R. sphaeroides* RV [209], and heterotrophic consortium and *R. sphaeroides* N7 [210]. However, the implementation of this integrated dark fermentation-photo-fermentation system is still hindered by the great differences in growth rate and acid tolerance between anaerobic bacteria and PNSB [211]. Alternatively, dark fermentation and photo-fermentation can be performed sequentially in separated reactors. In this process configuration, dark fermentation effluent containing VFAs is fed, after some adjustments such as dilution and neutralization [204], into photo-fermentation reactor to allow the conversion of VFAs to hydrogen by PNSB. This sequential process is generally easier to operate and control compared with the co-cultivation system as dark fermentation and photo-fermentation are operated separately. Recently, the sequential dark fermentation-photo-fermentation process was tested at a pilot scale using corn stover hydrolysate as a substrate in 11 m^3 reactor (3 m^3 for dark fermentation and 8 m^3 for photo-fermentation). Results showed that 59.7 m^3/d of hydrogen was produced, of which 22.4 m^3/d was from dark fermentation and 37.3 m^3/d was from photo-fermentation [47]. This demonstrates clearly that the sequential dark fermentation-photo-fermentation process is more efficient in conversion of biomass into hydrogen, compared with a single-stage dark fermentation or photo-fermentation process.

7.2 Process for methane production

A process for fermentative production of methane is generally called AD. AD is a microbiologically mediated process, in which organic compounds are converted into methane and carbon dioxide in the absence of oxygen [212]. AD process consists of four sequential stages, hydrolysis, acidogenesis, acetogenesis, and methanogenesis, and involves several groups of microorganisms. The hydrolysis is a stage that macromolecules (protein, fat, carbohydrate) are degraded to water soluble monomers (amino acids, fatty acids, and sugars). These monomers are then fermented to VFAs (acetic, propionic, lactic, butyric, and valeric acids) during the acidogenesis stage. The fermentation products after acidogenesis are subsequently converted into acetic acid, carbon dioxide, and hydrogen in the acetogenesis stage before acetic acid and hydrogen are consumed to produce methane [213]. AD process have been used to produce methane from a wide variety of lignocellulosic biomass, e.g., corn stover, barley straw, rice straw, wheat straw, sugarcane bagasse, and yard waste [200, 214]. Biochemical methane potential (BMP) of a selected biomass with a formula $C_aH_bO_c$ can be estimated using Buswell's equation (Eq. (18)), while Boyle's equation (Eq. (19)) is used to estimate BMP of biomass with a formula $C_aH_bO_cN_dS_e$, where a, b, c, d, e is the molar fraction of C, H, O, N, S, respectively. It should be noted that Eqs. (18) and (19) are used assuming the total stoichiometric conversion of organic matter into methane and carbon dioxide [215]. Using cellulose ($C_6H_{10}O_5$) as an example, BMP estimated using Eq. (18) is 415 mL/g-VS:

$$\mathrm{BMP} = \frac{\left(\frac{a}{2} + \frac{b}{8} - \frac{c}{4}\right)}{(12a + b + 16c)} \times 22,400 \tag{18}$$

$$\mathrm{BMP} = \frac{\left(\frac{a}{2} + \frac{b}{8} - \frac{c}{4} - \frac{3d}{8} - \frac{e}{4}\right)}{(12a + b + 16c + 14d + 32e)} \times 22,400 \tag{19}$$

Alternatively, organic fraction composition of biomass can be used to estimate the theoretical methane production using Eq. (20) [216]:

$$\mathrm{BMP} = (415 \times \%\mathrm{carbohydrate}) + (496 \times \%\mathrm{protein}) + (1014 \times \%\mathrm{lipid}) \tag{20}$$

AD process can be divided, based on the percentage of total solids (TS) in the system, into liquid-AD (L-AD) and solid-state AD (SS-AD). Although the criteria for this classification is not clear, it is generally accepted that systems containing less than 15% TS are called L-SD and those containing 15% TS or higher are called SS-AD. While L-AD is a traditional process being used extensively for waste treatment, SS-AD is relatively new, being developed in the past decades for municipal solid waste treatment [217]. Comparing between the two, SS-AD has many advantages over L-AD, including a smaller reactor volume, thus higher volumetric productivity of methane, higher organic loading rate, lower water consumption, lower energy input for operation (heating and mixing), and no problems of floating and stratification of fats [218]. However, due to a relatively high TS content of the system, limitation of mass and heat transfers can occur during the process, leading to a low fermentation yield. The use of SS-AD on wheat straw, corn stover, switch grass, and grass silage was reported to produce 55–197 L-CH_4/kg-volatile solids [219], while methane production of 45–290 L/kg-volatile solids were obtained from rice straw, corn straw, wheat straw, and yard waste [200].

8. Bioconversion process for lignocellulosic materials to bio-hydrogen and methane

Based on average composition of lignocellulose, 35–50% cellulose, 20–35% hemicellulose, and 10–25% lignin [220], bioconversion processes for cellulose into hydrogen and methane through dark fermentation, photo-fermentation, sequential dark fermentation-photo-fermentation, and AD are presented (**Figure 4**). Starting with 1000 kg of lignocellulosic biomass containing 35–50% cellulose, 193.4–276.3 m^3 of hydrogen is obtained by dark fermentation, 580–828.8 m^3 of hydrogen is obtained by photo-fermentation and a sequential dark fermentation-photo-fermentation, and 145.0–207.2 m^3 of methane is obtained by AD.

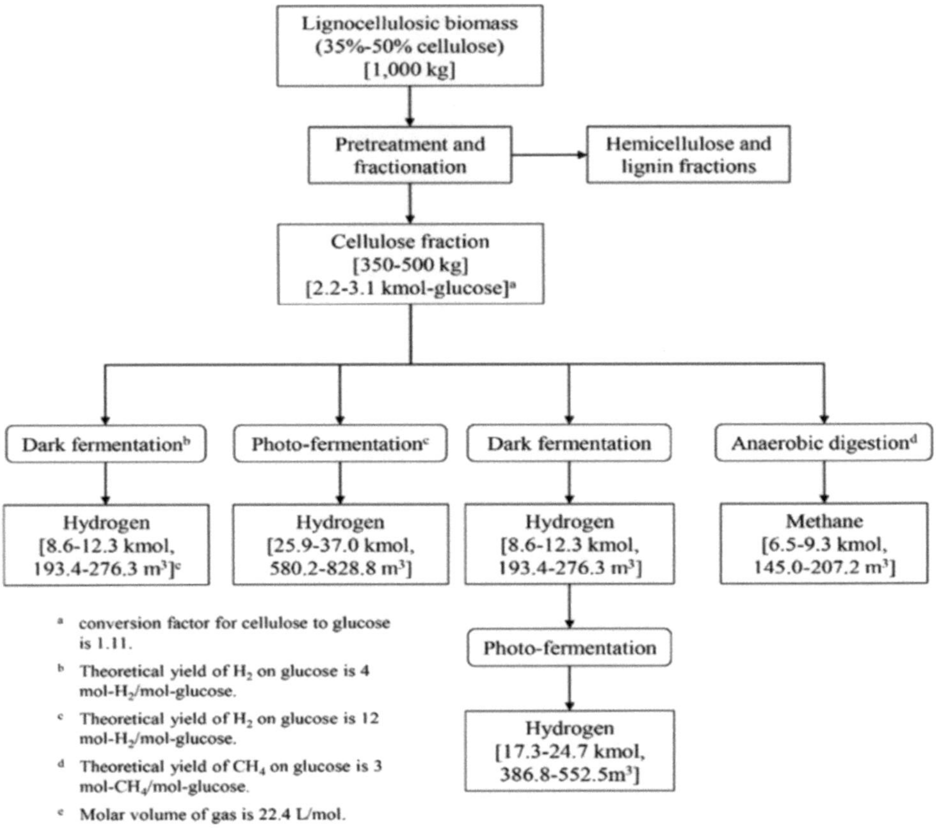

Figure 4.
Theoretical hydrogen and methane yields obtained from dark fermentation, photo-fermentation, sequential dark fermentation-photo-fermentation, and anaerobic digestion for hydrogen and methane production from lignocellulosic biomass.

9. Conclusion

Lignocellulosic materials are the promising substrate for bio-hydrogen and methane production. The main compositions of lignocellulosic materials are cellulose, hemicellulose, and lignin. In order to obtain the underlying monosugar, the appropriate pretreatment methods are recommended to hydrolyze hemicellulose or to remove lignin. Physical pretreatment can be used to increase the accessible surface area of lignocellulosic materials prior to the subsequent hydrolysis. Chemical hydrolysis by means of dilute acid pretreatment is an effective method to hydrolyze both hemicellulose and cellulose in lignocellulosic materials. Though, this method can slightly degrade lignin. Lignin can be effectively removed by alkaline pretreatment. Biological pretreatment by microorganisms and enzymes can be used to pretreat the lignocellulosic materials before enzymatic hydrolysis. Combined physical, chemical, and biological pretreatment process is more effective than a sole process. Bio-hydrogen and methane production process is greatly affected by the environmental factors. The ranges of these factors mainly depend on, but not limited to, the types of feedstocks and microorganisms in the process. Dark fermentation followed by photo-fermentation is more efficient in producing hydrogen from lignocellulosic materials than the single-stage fermentation. Methane production from lignocellulosic materials is an environmentally friendly process for producing bioenergy and managing the waste at the same time.

Acknowledgements

The authors gratefully acknowledge the TRF Senior Research Scholar (Grant No. RTA5980004) for the financial support to carry out research work on biofuels and value-added substances production from agricultural biomass.

Author details

Apilak Salakkam[1], Pensri Plangklang[1], Sureewan Sittijunda[2], Mallika Boonmee Kongkeitkajorn[1], Siriporn Lunprom[3] and Alissara Reungsang[1,3]*

1 Department of Biotechnology, Faculty of Technology, Khon Kaen University, Khon Kaen, Thailand

2 Faculty of Environment and Resource Studies, Mahidol University, Nakhon Pathom, Thailand

3 Research Group for Development of Microbial Hydrogen Production Process from Biomass, Khon Kaen University, Khon Kaen, Thailand

*Address all correspondence to: alissara@kku.ac.th

References

[1] World Bank. World Bank Open Data: Energy & Mining [Internet]. Available from: https://data.worldbank.org/ [Accessed: Dec 16, 2018]

[2] Ladanai S, Vinterbäck J. Global Potential of Sustainable Biomass for Energy [Internet]. 2009. Available from: https://pub.epsilon.slu.se/4523/1/ladanai_et_al_100211.pdf [Accessed: Dec 16, 2018]

[3] World Energy Council. World Energy Resources [Internet]. 2016. Available from: https://www.worldenergy.org/wp-content/uploads/2017/03/WEResources_Bioenergy_2016.pdf [Accessed: Dec 16, 2018]

[4] NZ Wood. What is the difference between hardwood and softwood? [Internet]. 2018. Available from: http://www.nzwood.co.nz/faqs/what-is-the-difference-between-hardwood-and-softwood/ [Accessed: Dec 16, 2018]

[5] Tye YY, Lee KT, Abdullah WNW, Leh CP. The world availability of non-wood lignocellulosic biomass for the production of cellulosic ethanol and potential pretreatments for the enhancement of enzymatic saccharification. Renewable and Sustainable Energy Reviews. 2016; **60**:155-172. DOI: 10.1016/j.rser.2016.01.072

[6] Barakat A, de Vries H, Rouau X. Dry fractionation process as an important step in current and future lignocellulose biorefineries: A review. Bioresource Technology. 2013;**134**:362-373. DOI: 10.1016/j.biortech.2013.01.169

[7] Menon V, Rao M. Trends in bioconversion of lignocellulose: Biofuels, platform chemicals & biorefinery concept. Progress in Energy and Combustion Science. 2012;**38**: 522-550. DOI: 10.1016/j.pecs.2012.02.002

[8] Payne CM, Knott BC, Mayes HB, Hansson H, Himmel ME, Sandgren M, et al. Fungal Cellulases. Chemical Reviews. 2015;**115**:1308-1448. DOI: 10.1021/cr500351c

[9] Gírio FM, Fonseca C, Carvalheiro F, Duarte LC, Marques S, Bogel-Łukasik R. Hemicelluloses for fuel ethanol: A review. Bioresource Technology. 2010; **101**:4775-4800. DOI: 10.1016/j.biortech.2010.01.088

[10] Pettersen RC. The chemical composition of wood. In: Rowell R, editor. The Chemistry of Solid Wood. Advances in Chemistry: American Chemical Society; Vol. 207. 1984. pp. 57-126. DOI: 10.1021/ba-1984-0207.ch002

[11] Peng F, Peng P, Xu F, Sun RC. Fractional purification and bioconversion of hemicelluloses. Biotechnology Advances. 2012;**30**: 879-903. DOI: 10.1016/j.biotechadv.2012.01.018

[12] Laurichesse S, Avérous L. Chemical modification of lignin: Towards biobased polymer. Progress in Polymer Science. 2014;**39**:1266-1290. DOI: 10.1016/j.progpolymsci.2013.11.004

[13] Bomble YJ, Lin CY, Amore A, Wei H, Holwerda EK, Ciesielski PN, et al. Lignocellulose deconstruction in the biosphere. Current Opinion in Chemical Biology. 2017;**41**:61-70. DOI: 10.1016/j.cbpa.2017.10.013

[14] Taha M, Foda M, Shahsavari E, Aburto-Medina A, Adetutu E, Ball A. Commercial feasibility of lignocellulose biodegradation: Possibilities and challenges. Current Opinion in Biotechnology. 2016;**38**:190-197. DOI: 10.1016/j.copbio.2016.02.012

[15] Brodeur G, Yau E, Badal K, Collier J, Ramachandran KB, Ramakrishnan S.

Chemical and physicochemical pretreatment of lignocellulosic biomass: A review. Enzyme Research. 2011;**17**: 1-18. DOI: 10.4061/2011/787532

[16] Sun Y, Cheng J. Hydrolysis of lignocellulosic materials for ethanol production: A review. Bioresource Technology. 2002;**83**:1-11. DOI: 10.1016/S0960-8524(01)00212-7

[17] Shahzadi T, Mehmood S, Irshad M, Anwar Z, Afroz A, Zeeshan N, et al. Advances in lignocellulosic biotechnology: A brief review on lignocellulosic biomass and cellulases. Advances in Bioscience and Biotechnology. 2014;**5**:6. DOI: 10.4236/abb.2014.53031

[18] Aguilar-Reynosa A, Romaní A, Rodríguez-Jasso R, Aguilar CN, Garrote G, Ruiz HA. Microwave heating processing as alternative of pretreatment in second-generation biorefinery: An overview. Energy Conversion and Management. 2017;**136**: 50-65. DOI: 10.1016/j.enconman.2017.01.004

[19] Braconnot H. Hydrolysis of cellulose into sugar. Gillbert's Annalen der Physiks. 1819;**63**:348

[20] Fangkum A, Reungsang A. Biohydrogen production from sugarcane bagasse hydrolysate by elephant dung: Effects of initial pH and substrate concentration. International Journal of Hydrogen Energy. 2011;**36**: 8687-8696. DOI: 10.1016/j.ijhydene.2010.05.119

[21] Kumar AK, Sharma S. Recent updates on different methods of pretreatment of lignocellulosic feedstocks: A review. Bioresources and Bioprocessing. 2017;**4**:1-19. DOI: 10.1186/s40643-017-0137-9

[22] Hassan SS, Williams GA, Jaiswal AK. Emerging technologies for the pretreatment of lignocellulosic biomass. Bioresource Technology. 2018;**262**: 310-318. DOI: 10.1016/j.biortech.2018.04.099

[23] Kumar P, Barrett DM, Delwiche MJ, Stroeve P. Methods for pretreatment of lignocellulosic biomass for efficient hydrolysis and biofuel production. Industrial & Engineering Chemistry Research. 2009;**48**:3713-3729. DOI: 10.1021/ie801542g

[24] Seidl PR, Goulart AK. Pretreatment processes for lignocellulosic biomass conversion to biofuels and bioproducts. Current Opinion in Green and Sustainable Chemistry. 2016;**2**: 48-53. DOI: 10.1016/j.cogsc.2016.09.003

[25] Elgharbawy AA, Alam MZ, Moniruzzaman M, Goto M. Ionic liquid pretreatment as emerging approaches for enhanced enzymatic hydrolysis of lignocellulosic biomass. Biochemical Engineering Journal. 2016;**109**:252-267. DOI: 10.1016/j.bej.2016.01.021

[26] Zhang K, Pei Z, Wang D. Organic solvent pretreatment of lignocellulosic biomass for biofuels and biochemicals: A review. Bioresource Technology. 2016;**199**:21-33. DOI: 10.1016/j.biortech.2015.08.102

[27] Thungklin P, Sittijunda S, Reungsang A. Sequential fermentation of hydrogen and methane from steam-exploded sugarcane bagasse hydrolysate. International Journal of Hydrogen Energy. 2018;**43**:9924-9934. DOI: 10.1016/j.ijhydene.2018.04.056

[28] Sun S, Sun S, Cao X, Sun R. The role of pretreatment in improving the enzymatic hydrolysis of lignocellulosic materials. Bioresource Technology. 2016;**199**:49-58. DOI: 10.1016/j.biortech.2015.08.061

[29] Rabemanolontsoa H, Saka S. Various pretreatments of lignocellulosics. Bioresource

Technology. 2016;**199**:83-91. DOI: 10.1016/j.biortech.2015.08.029

[30] Bayer EA, Belaich J-P, Shoham Y, Lamed R. The cellulosomes: Multienzyme machines for degradation of plant cell wall polysaccharides. Annual Review of Microbiology. 2004; **58**:521-554. DOI: 10.1146/annurev.micro.57.030502.091022

[31] Maeda RN, Serpa VI, Rocha VAL, Mesquita RAA, Anna LMMS, de Castro AM, et al. Enzymatic hydrolysis of pretreated sugar cane bagasse using *Penicillium funiculosum* and *Trichoderma harzianum* cellulases. Process Biochemistry. 2011;**46**. DOI: 1196-1201. DO: 10.1016/j.procbio.2011.01.022

[32] Aggarwal NK, Goyal V, Saini A, Yadav A, Gupta R. Enzymatic saccharification of pretreated rice straw by cellulases from *Aspergillus niger* BK01. 3 Biotech. 2017;7:1-10. DOI: 10.1007/s13205-017-0755-0

[33] Tsai M-H, Lee W-C, Kuan W-C, Sirisansaneeyakul S, Savarajara A. Evaluation of different pretreatments of Napier grass for enzymatic saccharification and ethanol production. Energy Science & Engineering. 2018;**6**: 683-692. DOI: 10.1002/ese3.243

[34] Taniguchi M, Suzuki H, Watanabe D, Sakai K, Hoshino K, Tanaka T. Evaluation of pretreatment with *Pleurotus ostreatus* for enzymatic hydrolysis of rice straw. Journal of Bioscience and Bioengineering. 2005; **100**:637-643. DOI: 10.1263/jbb.100.637

[35] Yu J, Zhang J, He J, Liu Z, Yu Z. Combinations of mild physical or chemical pretreatment with biological pretreatment for enzymatic hydrolysis of rice hull. Bioresource Technology. 2009;**100**:903-908. DOI: 10.1016/j.biortech.2008.07.025

[36] Prapinagsorn W, Sittijunda S, Reungsang A. Co-digestion of napier grass and its silage with cow dung for methane production. Energies. 2017;**10**:1654. DOI: 10.3390/en10101654

[37] Keskin T, Hallenbeck PC. Hydrogen production from sugar industry wastes using single-stage photo-fermentation. Bioresource Technology. 2012;**112**: 131-136. DOI: 10.1016/j.biortech.2012.02.077

[38] Bhatia SK, Kim SH, Yoon JJ, Yang YH. Current status and strategies for second generation biofuel production using microbial systems. Energy Conversion and Management. 2017;**148**: 1142-1156. DOI: 10.1016/j.enconman.2017.06.073

[39] International Energy Agency. Biofuels for Transportation [Internet]. 2018. Available from: https://www.iea.org/tcep/transport/biofuels/ [Accessed: Nov 18, 2018]

[40] Kapdan IK, Kargi F. Bio-hydrogen production from waste materials. Enzyme and Microbial Technology. 2006;**38**:569-582. DOI: 10.1016/j.enzmictec.2005.09.015

[41] Zajic JE, Kosaric N, Brosseau JD. Microbial production of hydrogen. In: Advances in Biochemical Engineering Vol. 9. Berlin/Heidelberg: Springer; 1978. p. 57-109. DOI: 10.1007/BFb0048091.

[42] Łukajtis R, Hołowacz I, Kucharska K, Glinka M, Rybarczyk P, Przyjazny A, et al. Hydrogen production from biomass using dark fermentation. Renewable and Sustainable Energy Reviews. 2018;**91**:665-694. DOI: 10.1016/j.rser.2018.04.043

[43] Momirlan M, Veziroglu TN. Current status of hydrogen energy. Renewable and Sustainable Energy Reviews. 2002;**6**:141-179. DOI: 10.1016/S1364-0321(02)00004-7

[44] Hallenbeck PC, Ghosh D. Advances in fermentative biohydrogen production: The way forward?. Trends in Biotechnology. 2009;**27**:287-297. DOI: 10.1016/j.tibech.2009.02.004

[45] Zhang Q, Wang Y, Zhang Z, Lee D-J, Zhou X, Jing Y, et al. Photo-fermentative hydrogen production from crop residue: A mini review. Bioresource Technology. 2017;**229**:222-230. DOI: 10.1016/ j.biortech.2017.01.008

[46] Zhang Z, Li Y, Zhang H, He C, Zhang Q. Potential use and the energy conversion efficiency analysis of fermentation effluents from photo and dark fermentative bio-hydrogen production. Bioresource Technology. 2017;**245**:884-889. DOI: 10.1016/j. biortech.2017.09.037

[47] Zhang Q, Zhang Z, Wang Y, Lee D-J, Li G, Zhou X, et al. Sequential dark and photo-fermentation hydrogen production from hydrolyzed corn stover: A pilot test using 11 m3 reactor. Bioresource Technology. 2018;**253**: 382-386. DOI: 10.1016/j. biortech.2018.01.017

[48] Elsamadony M, Tawfik A. Maximization of hydrogen fermentative process from delignified water hyacinth using sodium chlorite. Energy Conversion and Management. 2018;**157**: 257-265. DOI: 10.1016/j. enconman.2017.12.013

[49] Arreola-Vargas J, Flores-Larios A, González-Álvarez V, Corona-González RI, Méndez-Acosta HO. Single and two-stage anaerobic digestion for hydrogen and methane production from acid and enzymatic hydrolysates of Agave tequilana bagasse. International Journal of Hydrogen Energy. 2016;**41**:897-904. DOI: 10.1016/j.ijhydene.2015.11.016

[50] Contreras-Dávila CA, Méndez-Acosta HO, Arellano-García L, Alatriste-Mondragón F, Razo-Flores E. Continuous hydrogen production from enzymatic hydrolysate of Agave tequilana bagasse: Effect of the organic loading rate and reactor configuration. Chemical Engineering Journal. 2017;**313**: 671-679. DOI: 10.1016/j.cej.2016.12.084

[51] Sangyoka S, Reungsang A, Lin CY. Optimization of biohydrogen production from sugarcane bagasse by mixed cultures using a statistical method. Sustainable Environment Research. 2016;**26**:235-242. DOI: 10.1016/j.serj.2016.05.001

[52] Rai PK, Singh SP, Asthana RK, Singh S. Biohydrogen production from sugarcane bagasse by integrating dark- and photo-fermentation. Bioresource Technology. 2014;**152**:140-146. DOI: 10.1016/j.biortech.2013.10.117

[53] Saratale GD, Saratale RG, Kim SH, Kumar G. Screening and optimization of pretreatments in the preparation of sugarcane bagasse feedstock for biohydrogen production and process optimization. International Journal of Hydrogen Energy. 2018;**43**: 11470-11483. DOI: 10.1016/ j. ijhydene.2018.01.187

[54] Zhang SC, Lai QH, Lu Y, Liu ZD, Wang TM, Zhang C, et al. Enhanced biohydrogen production from corn stover by the combination of *Clostridium cellulolyticum* and hydrogen fermentation bacteria. Journal of Bioscience and Bioengineering. 2016; **122**:482-487. DOI: 10.1016/j. biosc.2016.03.014

[55] Xu J-F, Ren N, Su D-X, Qiu J. Bio-hydrogen production from acetic acid steam-exploded corn straws by simultaneous saccharification and fermentation with *Ethanoligenens harbinense* B49. International Journal of Energy Research. 2010;**34**:381-386. DOI: 10.1002/er.1659

[56] Li D, Chen H. Biological hydrogen production from steam-exploded straw by simultaneous saccharification and

fermentation. International Journal of Hydrogen Energy. 2007;**32**:1742-1748. DOI: 10.1016/j.ijhydene.2006.12.011

[57] O-Thong S, Prasertsan P, Karakashev D, Angelidaki I. High-rate continuous hydrogen production by *Thermoanaerobacterium thermosaccharolyticum* PSU-2 immobilized on heat-pretreated methanogenic granules. International Journal of Hydrogen Energy. 2008;**33**: 6498-6508. DOI: 10.1016/j.ijhydene.2008.07.060

[58] Sitthikitpanya S, Reungsang A, Prasertsan P, Khanal SK. Two-stage thermophilic bio-hydrogen and methane production from oil palm trunk hydrolysate using *Thermoanaerobacterium thermosaccharolyticum* KKU19. International Journal of Hydrogen Energy. 2017;**42**:28222-28232. DOI: 10.1016/ j.ijhydene.2017.09.136

[59] Islam MS, Zhang C, Sui KY, Guo C, Liu CZ. Coproduction of hydrogen and volatile fatty acid via thermophilic fermentation of sweet sorghum stalk from co-culture of *Clostridium thermocellum* and *Clostridium thermosaccharolyticum*. International Journal of Hydrogen Energy. 2017;**42**: 830-837. DOI: 10.1016/j.ijhydene.2016.09.117

[60] Cheng JR, Zhu MJ. Biohydrogen production from pretreated lignocellulose by *Clostridium thermocellum*. Biotechnology and Bioprocess Engineering. 2016;**21**:87-94. DOI: 10.1007/s12257-015-0642-7

[61] Zhang K, Ren NQ, Wang AJ. Fermentative hydrogen production from corn stover hydrolyzate by two typical seed sludges: Effect of temperature. International Journal of Hydrogen Energy. 2015;**40**: 3838-3848. DOI: 10.1016/j.ijhydene.2015.01.120

[62] Khamtib S, Plangklang P, Reungsang A. Optimization of fermentative hydrogen production from hydrolysate of microwave assisted sulfuric acid pretreated oil palm trunk by hot spring enriched culture. International Journal of Hydrogen Energy. 2011;**36**:14204-14216. DOI: 10.1016/j.ijhydene.2011.05.117

[63] Cao G, Ren N, Wang A, Lee DJ, Guo W, Liu B, et al. Acid hydrolysis of corn stover for biohydrogen production using *Thermoanaerobacterium thermosaccharolyticum* W16. International Journal of Hydrogen Energy. 2009;**34**:7182-7188. DOI: 10.1016/j.ijhydene.2009.07.009

[64] Ren NQ, Cao GL, Guo WQ, Wang AJ, Zhu YH, Liu BF, et al. Biological hydrogen production from corn stover by moderately thermophile *Thermoanaerobacterium thermosaccharolyticum* W16. International Journal of Hydrogen Energy. 2010;**35**:2708-2712. DOI: 10.1016/j.ijhydene.2009.04.044

[65] Islam MS, Guo C, Liu CZ. Enhanced hydrogen and volatile fatty acid production from sweet sorghum stalks by two-steps dark fermentation with dilute acid treatment in between. International Journal of Hydrogen Energy. 2018;**43**:659-666. DOI: 10.1016/ j.ijhydene.2017.11.059

[66] Sitthikitpanya S, Reungsang A, Prasertsan P. Two-stage thermophilic bio-hydrogen and methane production from lime-pretreated oil palm trunk by simultaneous saccharification and fermentation. International Journal of Hydrogen Energy. 2018;**43**:4284-4293. DOI: 10.1016/j.ijhydene.2018.01.063

[67] Zhao L, Cao GL, Wang AJ, Guo WQ, Ren HY, Ren NQ. Simultaneous saccharification and fermentation of fungal pretreated cornstalk for hydrogen production using *Thermoanaerobacterium*

thermosaccharolyticum W16. Bioresource Technology. 2013;**145**:103-107. DOI: 10.1016/j.bopttech.2013.01.144

[68] Khamtib S, Reungsang A. Co-digestion of oil palm trunk hydrolysate with slaughterhouse wastewater for thermophilic bio-hydrogen production by *Thermoanaerobacterium thermosaccharolyticm* KKU19. International Journal of Hydrogen Energy. 2014;**39**:6872-6880. DOI: 10.1016/j.ijhydene.2014.02.073

[69] Prapinagsorn W, Sittijunda S, Reungsang A, Prapinagsorn W, Sittijunda S, Reungsang A. Co-digestion of napier grass and its silage with cow dung for bio-hydrogen and methane production by two-stage anaerobic digestion process. Energies. 2017;**11**:47. DOI: 10.3390/en11010047

[70] Lopez-Hidalgo AM, Alvarado-Cuevas ZD, De Leon-Rodriguez A. Biohydrogen production from mixtures of agro-industrial wastes: Chemometric analysis, optimization and scaling up. Energy. 2018;**159**:32-41. DOI: 10.1016/j.energy.2018.06.124

[71] Mirza S, Qazi J, Zhao Q, Chen S. Photo-biohydrogen production potential of *Rhodobacter capsulatus*-PK from wheat straw. Biotechnology for Biofuels. 2013;**6**:144. DOI: 10.1186/1754-6834-6-144

[72] Jiang D, Ge X, Zhang T, Liu H, Zhang Q. Photo-fermentative hydrogen production from enzymatic hydrolysate of corn stalk pith with a photosynthetic consortium. International Journal of Hydrogen Energy. 2016;**41**:16778-16785. DOI: 10.1016/j.ijhydene.2016.07.129

[73] Su H, Cheng J, Zhou J, Song W, Cen K. Hydrogen production from water hyacinth through dark- and photo-fermentation. International Journal of Hydrogen Energy. 2010;**35**:8929-8937. DOI: 10.1016/j.ijhydene.2010.06.035

[74] Yang H, Shi B, Ma H, Guo L. Enhanced hydrogen production from cornstalk by dark- and photo-fermentation with diluted alkali-cellulase two-step hydrolysis. International Journal of Hydrogen Energy. 2015;**40**:12193-12200. DOI: 10.1016/j.ijhydene.2015.07.062

[75] Yang H, Guo L, Liu F. Enhanced bio-hydrogen production from corncob by a two-step process: Dark- and photo-fermentation. Bioresource Technology. 2010;**101**:2049-2052. DOI: 10.1016/j.biortech.2009.10.078

[76] Li Y, Khanal SK. Bioenergy : Principles and Applications. Hoboken, New Jersey: Jhon Wiley & Sons, Inc.; 2017. 569 p. ISBN: 978-1-118-56831-6

[77] Korakianitis T, Namasivayam AM, Crookes RJ. Natural-gas fueled spark-ignition (SI) and compression-ignition (CI) engine performance and emissions. Progress in Energy and Combustion Science. 2011;**37**:89-112. DOI: 10.1016/j.pecs.2010.04.002

[78] Gerardi MH. The Microbiology of Anaerobic Digesters. Hoboken, New Jersy: John Wiley & Sons, Inc.; 2003. 177 p. DOI: 10.1002/0471468967

[79] Liu CM, Wachemo AC, Yuan HR, Zou DX, Liu YP, Zhang L, et al. Evaluation of methane yield using acidogenic effluent of NaOH pretreated corn stover in anaerobic digestion. Renewable Energy. 2018;**116**:224-233. DOI: 10.1016/j.renene.2017.07.001

[80] Eskicioglu C, Monlau F, Barakat A, Ferrer I, Kaparaju P, Trably E, et al. Assessment of hydrothermal pretreatment of various lignocellulosic biomass with CO_2 catalyst for enhanced methane and hydrogen production. Water Research. 2017;**120**:32-42. DOI: 10.1016/j.watres.2017.04.068

[81] Narra M, Macwan K, Vyas B, Harijan MR, Shah D, Balasubramanian V, et al.

A bio-refinery concept for production of bio-methane and bio-ethanol from nitric acid pre-treated corncob and recovery of a high value fuel from a waste stream. Renewable Energy. 2018; **127**:1-10. DOI: 10.1016/j.renene. 2018.04.044

[82] Kasprzycka A, Lalak-Kańczugowska J, Tys J. Flammulina velutipes treatment of non-sterile tall wheat grass for enhancing biodegradability and methane production. Bioresource Technology. 2018;**263**:660-664. DOI: 10.1016/j.biortech.2018.05.024

[83] Hao X, Hu Y, Cao D. Destroying lignocellulosic matters for enhancing methane production from excess sludge. Environmental Technology. 2016;**37**: 623-629. DOI: 10.1080/09593330. 2015.1075600

[84] Hassan M, Umar M, Mamat T, Muhayodin F, Talha Z, Mehryar E, et al. Methane enhancement through sequential thermochemical and sonication pretreatment for corn stover with anaerobic sludge. Energy and Fuels. 2017;**31**:6145-6153. DOI: 10.1021/ acs.energyfuels.7b00478

[85] Hassan M, Ding W, Umar M, Hei K, Bi J, Shi Z. Methane enhancement and asynchronism minimization through co-digestion of goose manure and NaOH solubilized corn stover with waste activated sludge. Energy. 2017;**118**: 1256-1263. DOI: 10.1016/j.energy. 2016.11.007

[86] Zhang Z, Zhang G, Li W, Li C, Xu G. Enhanced biogas production from sorghum stem by co-digestion with cow manure. International Journal of Hydrogen Energy. 2018;**41**:9153-9158. DOI: 10.1016/j.ijhydene.2016.02.042

[87] Yao Y, Bergeron AD, Davaritouchaee M. Methane recovery from anaerobic digestion of urea-pretreated wheat straw. Renewable Energy. 2018;**115**:139-148. DOI: 10.1016/ j.renene.2017.08.038

[88] Peng X, Nges IA, Liu J. Improving methane production from wheat straw by digestate liquor recirculation in continuous stirred tank processes. Renewable Energy. 2016:8512-8518. DOI: 10.1016/j.renene.2015.06.023

[89] Mustafa AM, Poulsen TG, Sheng K. Fungal pretreatment of rice straw with *Pleurotus ostreatus* and *Trichoderma reesei* to enhance methane production under solid-state anaerobic digestion. Applied Energy. 2016;**180**:661-671. DOI: 10.1016/j.apenergy.2016.07.135

[90] Xu W, Fu S, Yang Z, Lu J, Guo R. Improved methane production from corn straw by microaerobic pretreatment with a pure bacteria system. Bioresource Technology. 2018; **259**:18-23. DOI: 10.1016/j.biortech. 2018.02.046

[91] Fu SF, Wang F, Yuan XZ, Yang ZM, Luo SJ, Wang CS, et al. The thermophilic (55°C) microaerobic pretreatment of corn straw for anaerobic digestion. Bioresource Technology. 2015;**175**:203-208. DOI: 10.1016/j.biortech.2014.10.072

[92] Andriamanohiarisoamanana FJ, Matsunami N, Yamashiro T, Iwasaki M, Ihara I, Umetsu K. High-solids anaerobic mono-digestion of riverbank grass under thermophilic conditions. Journal of Environmental Sciences. 2017;**52**:29-38. DOI: 10.1016/j. jes.2016.05.005

[93] Khor WC, Vervaeren H, Rabaey K. Combined extrusion and alkali pretreatment improves grass storage towards fermentation and anaerobic digestion. Biomass and Bioenergy. 2018; **119**:121-127. DOI: 10.1016/j.biombioe. 2018.09.003

[94] Alexandropoulou M, Antonopoulou G, Fragkou E, Ntaikou I, Lyberatos G.

Fungal pretreatment of willow sawdust and its combination with alkaline treatment for enhancing biogas production. Journal of Environmental Management. 2017;**203**:704-713. DOI: 10.1016/j.jenvman.2016.04.006

[95] Bolado-Rodríguez S, Toquero C, Martín-Juárez J, Travaini R, García-Encina PA. Effect of thermal, acid, alkaline and alkaline-peroxide pretreatments on the biochemical methane potential and kinetics of the anaerobic digestion of wheat straw and sugarcane bagasse. Bioresource Technology. 2016;**201**:182-190. DOI: 10.1016/j.biortech.2015.11.047

[96] Cao W, Sun C, Li X, Qiu J, Liu R. Methane production enhancement from products of alkaline hydrogen peroxide pretreated sweet sorghum bagasse. RSC Advances. 2017;**7**:5701-5707. DOI: 10.1039/c6ra25798d

[97] Benito Martin PC, Schlienz M, Greger M. Production of bio-hydrogen and methane during semi-continuous digestion of maize silage in a two-stage system. International Journal of Hydrogen Energy. 2017;**42**:5768-5779. DOI: 10.1016/j.ijhydene.2017.01.020

[98] Kumari S, Das D. Biohythane production from sugarcane bagasse and water hyacinth: A way towards promising green energy production. Journal of Cleaner Production. 2019; **207**:689-701. DOI: 10.1016/j.jclepro.2018.10.050

[99] Montiel Corona V, Razo-Flores E. Continuous hydrogen and methane production from *Agave tequilana* bagasse hydrolysate by sequential process to maximize energy recovery efficiency. Bioresource Technology. 2018;**249**:334-341. DOI: 10.1016/j.biortech.2017.10.032

[100] Nualsri C, Reungsang A, Plangklang P. Biochemical hydrogen and methane potential of sugarcane syrup using a two-stage anaerobic fermentation process. Industrial Crops and Products. 2016;**82**:88-99. DOI: 10.1016/j.indcrop.2015.12.002

[101] Li CL, Fang HHP. Fermentative hydrogen production from wastewater and solid wastes by mixed cultures. Critical Reviews in Environmental Science and Technology. 2007;**37**:1-39. DOI: 10.1080/10643380600729071

[102] Ghimire A, Frunzo L, Pirozzi F, Trably E, Escudie R, Lens PNL, et al. A review on dark fermentative biohydrogen production from organic biomass: Process parameters and use of by-products. Applied Energy. 2015;**144**: 73-95. DOI: 10.1016/j.apenergy.2015.01.045

[103] De Gioannis G, Muntoni A, Polettini A, Pomi R. A review of dark fermentative hydrogen production from biodegradable municipal waste fractions. Waste Management. 2013;**33**: 1345-1361. DOI: 10.1016/j.wasman.2013.02.019

[104] Chaganti SR, Kim DH, Lalman JA. Impact of oleic acid on the fermentation of glucose and xylose mixtures to hydrogen and other byproducts. Renewable Energy. 2012;**42**:60-65. DOI: 10.1016/j.renene.2011.09.015

[105] Cheng CL, Lo YC, Lee KS, Lee DJ, Lin CY, Chang JS. Biohydrogen production from lignocellulosic feedstock. Bioresource Technology. 2011;**102**:8514-8523. DOI: 10.1016/j.biortech.2011.04.059

[106] Harun R, Danquah MK, Forde GM. Microalgal biomass as a fermentation feedstock for bioethanol production. Journal of Chemical Technology and Biotechnology. 2010;**85**:199-203. DOI: 10.1002/jctb.2287

[107] Cheng CH, Du TB, Pi HC, Jang SM, Lin YH, Lee HT. Comparative study of lipid extraction from microalgae by

organic solvent and supercritical CO_2. Bioresource Technology. 2011;**102**: 10151-10153. DOI: 10.1016/j.biortech.2011.08.064

[108] Elbeshbishy E, Dhar BR, Nakhla G, Lee HS. A critical review on inhibition of dark biohydrogen fermentation. Renewable and Sustainable Energy Reviews. 2017;**79**:656-668. DOI: 10.1016/j.rser.2017.05.075

[109] Wang J, Wan W. Factors influencing fermentative hydrogen production: A review. International Journal of Hydrogen Energy. 2009;**34**: 799-811. DOI: 10.1016/j.ijhydene.2008.11.015

[110] Logan BE, Oh SE, Kim IS, Van Ginkel S. Biological hydrogen production measured in batch anaerobic respirometers. Environmental Science & Technology. 2002;**36**:2530-2535. DOI: 10.1021/es015783i

[111] Oh SE, Van Ginkel S, Logan BE. The relative effectiveness of pH control and heat treatment for enhancing biohydrogen gas production. Environmental Science & Technology. 2003;**37**:5186-5190. DOI: 10.1021/es034291y

[112] Zhang T, Liu H, Fang HHP. Biohydrogen production from starch in wastewater under thermophilic condition. Journal of Environmental Management. 2003;**69**:149-156. DOI: 10.1016/S0301-4797(03)00141-5

[113] Bisaillon A, Turcot J, Hallenbeck PC. The effect of nutrient limitation on hydrogen production by batch cultures of *Escherichia coli*. International Journal of Hydrogen Energy. 2006;**31**:1504-1508. DOI: 10.1016/j.ijhydene.2006.06.016

[114] Salerno MB, Park W, Zuo Y, Logan BE. Inhibition of biohydrogen production by ammonia. Water Research. 2006;**40**:1167-1172. DOI: 10.1016/j.watres.2006.01.024

[115] Ferchichi M, Crabbe E, Gil GH, Hintz W, Almadidy A. Influence of initial pH on hydrogen production from cheese whey. Journal of Biotechnology. 2005;**120**:402-409. DOI: 10.1016/j.jbiotec.2005.05.017

[116] Ueno Y, Haruta S, Ishii M, Igarashi Y. Characterization of a microorganism isolated from the effluent of hydrogen fermentation by microflora. Journal of Bioscience and Bioengineering. 2001;**92**: 397-400. DOI: 10.1016/S1389-1723(01)80247-4

[117] Lay JJ, Fan KS, Hwang JI, Chang JI, Hsu PC. Factors affecting hydrogen production from food wastes by Clostridium-rich composts. Journal of Environmental Engineering. 2005;**131**: 595-602. DOI: 10.1061/(ASCE)0733-9372

[118] Argun H, Kargi F, Kapdan IK, Oztekin R. Biohydrogen production by dark fermentation of wheat powder solution: Effects of C/N and C/P ratio on hydrogen yield and formation rate. International Journal of Hydrogen Energy. 2008;**33**:1813-1819. DOI: 10.1016/j.ijhydene.2008.01.03

[119] O-Thong S, Prasertsan P, Intrasungkha N, Dhamwichukorn S, Birkeland NK. Optimization of simultaneous thermophilic fermentative hydrogen production and COD reduction from palm oil mill effluent by Thermoanaerobacterium-rich sludge. International Journal of Hydrogen Energy. 2008;**33**:1221-1231. DOI: 10.1016/j.ijhydene.2007.12.017

[120] Lay JJ. Biohydrogen generation by mesophilic anaerobic fermentation of microcrystalline cellulose. Biotechnology and Bioengineering. 2001;**74**:280-287. DOI: 10.1002/bit.1118

[121] Khanal SK. Anaerobic Biotechnology for Bioenergy Production: Principles and Applications. Iowa: Wiley- Blackwell; 2008. DOI: 10.1002/9780813804545

[122] Sinha P, Pandey A. An evaluative report and challenges for fermentative biohydrogen production. International Journal of Hydrogen Energy. 2011;**36**: 7460-7478. DOI: 10.1016/j.ijhydene. 2011.03.077

[123] Sotelo-Navarro PX, Poggi-Varaldo HM, Turpin-Marion SJ, Vázquez-Morillas A, Beltrán-Villavicencio M, Espinosa-Valdemar RM. Biohydrogen production from used diapers: Evaluation of effect of temperature and substrate conditioning. Waste Management Research. 2017;**35**:267-275. DOI: 10.1177/0734242X16677334

[124] Khanal SK, Chen WH, Li L, Sung S. Biological hydrogen production: Effects of pH and intermediate products. International Journal of Hydrogen Energy. 2004;**29**:1123-1131. DOI: 10.1016/j.ijhydene.2003.11.002

[125] Mohd Yasin NH, Rahman NA, Man HC, Mohd Yusoff MZ, Hassan MA. Microbial characterization of hydrogen-producing bacteria in fermented food waste at different pH values. International Journal of Hydrogen Energy. 2011;**36**:9571-9580. DOI: 10.1016/j.ijhydene.2011.05.048

[126] Cai ML, Liu JX, Wei YS. Enhanced biohydrogen production from sewage sludge with alkaline pretreatment. Environmental Science & Technology. 2004;**38**:3195-3202. DOI: 10.1021/es0349204

[127] Liu IC, Whang LM, Ren WJ, Lin PY. The effect of pH on the production of biohydrogen by clostridia: Thermodynamic and metabolic considerations. International Journal of Hydrogen Energy. 2011;**36**:439-449. DOI: 10.1016/j.ijhydene.2010.10.045

[128] Guo XM, Trably E, Latrille E, Carrère H, Steyer JP. Hydrogen production from agricultural waste by dark fermentation: A review. International Journal of Hydrogen Energy. 2010;**35**:10660-10673. DOI: 10.1016/j.ijhydene.2010.03.008

[129] Wang J, Wan W. Effect of Fe^{2+} concentration on fermentative hydrogen production by mixed cultures. International Journal of Hydrogen Energy. 2008;**33**:1215-1220. DOI: 10.1016/j.ijhydene.2007.12.044

[130] Wang J, Wan W. Influence of Ni^{2+} concentration on biohydrogen production. Bioresource Technology. 2008;**99**:8864-8868. DOI: 10.1016/j.biortech.2008.04.052

[131] Lee YJ, Miyahara T, Noike T. Effect of pH on microbial hydrogen fermentation. Journal of Chemical Technology and Biotechnology. 2002;**77**: 694-698. DOI: 10.1002/jctb.623

[132] Lin CY, Lay CH. A nutrient formulation for fermentative hydrogen production using anaerobic sewage sludge microflora. International Journal of Hydrogen Energy. 2005;**30**:285-292. DOI: 10.1016/j.ijhydene.2004.03.002

[133] Wang XJ, Ren NQ, Sheng Xiang W, Qian Guo W. Influence of gaseous end-products inhibition and nutrient limitations on the growth and hydrogen production by hydrogen-producing fermentative bacterial B49. International Journal of Hydrogen Energy. 2007;**32**:748-754. DOI: 10.1016/j.ijhydene.2006.08.003

[134] Jung KW, Kim DH, Kim SH, Shin HS. Bioreactor design for continuous dark fermentative hydrogen production. Bioresource Technology. 2011;**102**: 8612-8620. DOI: 10.1016/j.biortech. 2011.03.056

[135] Chong ML, Sabaratnam V, Shirai Y, Hassan MA. Biohydrogen production from biomass and industrial wastes by dark fermentation. International Journal of Hydrogen Energy. 2009;**34**: 3277-3287. DOI: 10.1016/j.ijhydene. 2009.02.010

[136] Nualsri C, Kongjan P, Reungsang A, Imai T. Effect of biogas sparging on the performance of bio-hydrogen reactor over a long-term operation. PLoS One. 2017;**12**:1-13. DOI: 10.1371/journal.pone.0171248

[137] Barbosa MJ, Rocha JMS, Tramper J, Wijffels RH. Acetate as a carbon source for hydrogen production by photosynthetic bacteria. Journal of Biotechnology. 2001;**85**:25-33. DOI: 10.1016/S0168-1656(00)00368-0

[138] Basak N, Das D. Photofermentative hydrogen production using purple non-sulfur bacteria *Rhodobacter sphaeroides* O.U.001 in an annular photobioreactor: A case study. Biomass and Bioenergy. 2009;**33**:911-919. DOI: 10.1016/j.biombioe.2009.02.007

[139] Wang YZ, Liao Q, Zhu X, Tian X, Zhang C. Characteristics of hydrogen production and substrate consumption of *Rhodopseudomonas palustris* CQK 01 in an immobilized-cell photobioreactor. Bioresource Technology. 2010;**101**: 4034-4041. DOI: 10.1016/j.biortech.2010.01.045

[140] Kim MS, Kim DH, Cha J. Culture conditions affecting H_2 production by phototrophic bacterium *Rhodobacter sphaeroides* KD131. International Journal of Hydrogen Energy. 2012;**37**: 14055-14061. DOI: 10.1016/j.ijhydene.2012.06.085

[141] Kim DH, Son H, Kim MS. Effect of substrate concentration on continuous Photo-fermentative hydrogen production from lactate using *Rhodobacter sphaeroides*. International Journal of Hydrogen Energy. 2012;**37**: 15483-15488. DOI: 10.1016/j.ijhydene.2012b.02.039

[142] Ozmihci S, Kargi F. Bio-hydrogen production by photo-fermentation of dark fermentation effluent with intermittent feeding and effluent removal. International Journal of Hydrogen Energy. 2010;**35**:6674-6680. DOI: 10.1016/j.ijhydene.2010.04.090

[143] Azbar N, Cetinkaya Dokgoz FT. The effect of dilution and L-malic acid addition on bio-hydrogen production with *Rhodopseudomonas palustris* from effluent of an acidogenic anaerobic reactor. International Journal of Hydrogen Energy. 2010;**35**:5028-5033. DOI: 10.1016/j.ijhydene.2009.10.044

[144] Sagnak R, Kargi F. Photo-fermentative hydrogen gas production from dark fermentation effluent of acid hydrolyzed wheat starch with periodic feeding. International Journal of Hydrogen Energy. 2011;**36**: 4348-4353. DOI: 10.1016/j.ijhydene.2011.01.033

[145] Ike A, Murakawa T, Kawaguchi H, Hirata K, Miyamoto K. Photoproduction of hydrogen from raw starch using a halophilic bacterial community. Journal of Bioscience and Bioengineering. 1999; **88**:72-77. DOI: 10.1016/S1389-1723(99)80179-0

[146] Yang H, Wang X, Zhang L, Guo L. Enhanced hydrogen production performance of *Rubrivivax gelatinosus* M002 using mixed carbon sources. International Journal of Hydrogen Energy. 2012;**37**:13296-13303. DOI: 10.1016/j.ijhydene.2012.06.101

[147] Zhu H, Ueda S, Asada Y, Miyake J. Hydrogen production as a novel process of wastewater treatment-studies on tofu wastewater with entrapped *R. sphaeroides* and mutagenesis. International Journal of Hydrogen Energy. 2002;**27**:1349-1357. DOI: 10.1016/S0360-3199(02)00118-0

[148] Eroğlu E, Gündüz U, Yücel M, Türker L, Eroğlu I. Photobiological hydrogen production from olive mill wastewater as sole substrate sources. International Journal of Hydrogen Energy. 2004;**29**:163-171. DOI: 10.1016/S0360-3199(03)00110-1

[149] Seifert K, Waligorska M, Laniecki M. Brewery wastewaters in photobiological hydrogen generation in presence of *Rhodobacter sphaeroides* O.U. 001. International Journal of Hydrogen Energy. 2010;**35**:4085-4091. DOI: 10.1016/j.ijhydene.2010.01.126

[150] Seifert K, Waligorska M, Laniecki M. Hydrogen generation in photobiological process from dairy wastewater. International Journal of Hydrogen Energy. 2010;**35**:9624-9629. DOI: 10.1016/j.ijhydene.2010.07.015

[151] Chen CY, Chang JS. Enhancing phototropic hydrogen production by solid-carrier assisted fermentation and internal optical-fiber illumination. Process Biochemistry. 2006;**41**: 2041-2049. DOI: 10.1016/j. procbio.2006.05.005

[152] Fang HHP, Liu H, Zhang T. Phototrophic hydrogen production from acetate and butyrate in wastewater. International Journal of Hydrogen Energy. 2005;**30**:785-793. DOI: 10.1016/ j.ijhydene.2004.12.010

[153] Chen CY, Liu CH, Lo YC, Chang JS. Perspectives on cultivation strategies and photobioreactor designs for photo-fermentative hydrogen production. Bioresource Technology. 2011;**102**: 8484-8492. DOI: 10.1016/j. biortech.2011.05.082

[154] Chen CY, Lu WB, Wu JF, Chang JS. Enhancing phototrophic hydrogen production of *Rhodopseudomonas palustris* via statistical experiment design. International Journal of Hydrogen Energy. 2007;**32**: 940-949. DOI: 10.1016/j.ijhydene. 2006.09.021

[155] Lee JZ, Klaus DM, Maness PC, Spear JR. The effect of butyrate concentration on hydrogen production via photo-fermentation for use in a Martian habitat resource recovery process. International Journal of Hydrogen Energy. 2007;**32**:3301-3307. DOI: 10.1016/j.ijhydene.2007.05.029

[156] Koku H, Eroğlu I, Gündüz U, Yücel M, Türker L. Aspects of the metabolism of hydrogen production by *Rhodobacter sphaeroides*. International Journal of Hydrogen Energy. 2002;**27**: 1315-1329. DOI: 10.1016/S0360-3199 (02)00127-1

[157] Chen CY, Lee CM, Chang JS. Feasibility study on bioreactor strategies for enhanced photohydrogen production from *Rhodopseudomonas palustris* WP3-5 using optical-fiber-assisted illumination systems. International Journal of Hydrogen Energy. 2006;**31**:2345-2355. DOI: 10.1016/j.ijhydene.2006.03.007

[158] Shi XY, Yu HQ. Optimization of glutamate concentration and pH for H_2 production from volatile fatty acids by *Rhodopseudomonas capsulata*. Letters in Applied Microbiology. 2005;**40**: 401-406. DOI: 10.1111/ j.1472-765X.2005.01700.x

[159] Cai J, Wang G. Hydrogen production by a marine photosynthetic bacterium, *Rhodovulum sulfidophilum* P5, isolated from a shrimp pond. International Journal of Hydrogen Energy. 2012;**37**:15070-15080. DOI: 10.1016/j.ijhydene.2012.07.130

[160] Tao Y, He Y, Wu Y, Liu F, Li X, Zong W, et al. Characteristics of a new photosynthetic bacterial strain for hydrogen production and its application in wastewater treatment. International Journal of Hydrogen Energy. 2008;**33**: 963-973. DOI: 10.1016/j.ijhydene.2007. 11.021

[161] Han H, Jia Q, Liu B, Yang H, Shen J. Fermentative hydrogen production from acetate using *Rhodobacter sphaeroides* RV. International Journal of Hydrogen Energy. 2013;**38**:10773-10778. DOI: 10.1016/j.ijhydene.2013.02.134

[162] Lazaro CZ, Varesche MBA, Silva EL. Effect of inoculum concentration, pH, light intensity and lighting regime on hydrogen production by phototrophic microbial consortium. Renewable Energy. 2015;**75**:1-7. DOI: 10.1016/j.renene.2014.09.034

[163] Laocharoen S, Reungsang A. Isolation, characterization and optimization of photo-hydrogen production conditions by newly isolated *Rhodobacter sphaeroides* KKU-PS5. International Journal of Hydrogen Energy. 2014;**39**: 10870-10882. DOI: 10.1016/j.ijhydene.2014.05.055

[164] Özgür E, Uyar B, Öztürk Y, Yücel M, Gündüz U, Eroğlu I. Biohydrogen production by *Rhodobacter capsulatus* on acetate at fluctuating temperatures. Resources. Conservation and Recycling. 2010;**54**:310-314. DOI: 10..1016/j.resconrec.2009.06.002

[165] Chen CY, Lee CM, Chang JS. Hydrogen production by indigenous photosynthetic bacterium *Rhodopseudomonas palustris* WP3-5 using optical fiber-illuminating photobioreactors. Biochemical Engineering Journal. 2006;**32**:33-42. DOI: 10.1016/j.bej.2006.08.015

[166] Akkerman I, Janssen M, Rocha J, Wijffels RH. Photobiological hydrogen production: Photochemical efficiency and bioreactor design. International Journal of Hydrogen Energy. 2002;**27**: 1195-1208. DOI: 10.1016/S0360-3199(02)00071-X

[167] Li X, Wang YH, Zhang SL, Chu J, Zhang M, Huang MZ, et al. Enhancement of phototrophic hydrogen production by *Rhodobacter sphaeroides* ZX-5 using a novel strategy-shaking and extra-light supplementation approach. International Journal of Hydrogen Energy. 2009;**34**:9677-9685. DOI: 10.1016/j.ijhydene.2009.10.020

[168] Kim NJ, Lee JK, Lee CG. Pigment reduction to improve photosynthetic productivity of *Rhodobacter sphaeroides*. Journal of Microbiology and Biotechnology. 2004;**14**:442-449

[169] Kim MS, Baek JS, Lee JK. Comparison of H_2 accumulation by *Rhodobacter sphaeroides* KD131 and its uptake hydrogenase and PHB synthase deficient mutant. International Journal of Hydrogen Energy. 2006;**31**:121-127. DOI: 10.1016/j.ijhydene.2004.10.023

[170] Kim MS, Kim DH, Cha J, Lee JK. Effect of carbon and nitrogen sources on photo-fermentative H_2 production associated with nitrogenase, uptake hydrogenase activity, and PHB accumulation in *Rhodobacter sphaeroides* KD131. Bioresource Technology. 2012; **116**:179-183. DOI: 10.1016/j.biortech.2012.04.011

[171] Argun H, Kargi F. Effects of light source, intensity and lighting regime on bio-hydrogen production from ground wheat starch by combined dark and photo-fermentations. International Journal of Hydrogen Energy. 2010;**35**: 1604-1612. DOI: 10.1016/j.ijhydene.2009.12.033

[172] Tian X, Liao Q, Liu W, Wang YZ, Zhu X, Li J, et al. Photo-hydrogen production rate of a PVA-boric acid gel granule containing immobilized photosynthetic bacteria cells. International Journal of Hydrogen Energy. 2009;**34**:4708-4717. DOI: 10.1016/j.ijhydene.2009.03.042

[173] Kawagoshi Y, Oki Y, Nakano I, Fujimoto A, Takahashi H. Biohydrogen production by isolated halotolerant photosynthetic bacteria using long-wavelength light-emitting diode (LW-LED). International Journal of Hydrogen Energy. 2010;**35**:13365-13369. DOI: 10.1016/j.ijhydene.2009.11.121

[174] Seefeldt LC, Yang ZY, Duval S, Dean DR. Nitrogenase reduction of

carbon-containing compounds. Biochimica et Biophysica Acta. 2013; **1827**:1102-1111. DOI: 10.1016/j.bbabio.2013.04.003

[175] Ludden PW, Roberts GP. The biochemistry and genetics of nitrogen fixation by photosynthetic bacteria. In: Blankenship RE, Madigan MT, Bauer CE, editors. Anoxygenic Photosynthetic Bacteria. Dordrecht: Kluwer Academic Publishers; 1995. pp. 929-947. DOI: 10.1007/0-306-47954-0

[176] Zhu H, Fang HHP, Zhang T, Beaudette LA. Effect of ferrous ion on photo heterotrophic hydrogen production by *Rhodobacter sphaeroides*. International Journal of Hydrogen Energy. 2007;**32**:4112-4118. DOI: 10.1016/j.ijhydene.2007.06.010

[177] Han H, Jia Q, Liu B, Yang H, Shen J. Optimization of photosynthetic hydrogen production from acetate by *Rhodobacter sphaeroides* RV. International Journal of Hydrogen Energy. 2013;**38**:12886-12890. DOI: 10.1016/j.ijhydene.2013.05.156

[178] Rai PK, Asthana RK, Singh SP. Optimization of photo-hydrogen production based on cheese whey spent medium. International Journal of Hydrogen Energy. 2014;**39**:7597-7603. DOI: 10.1016/j.ijhydene.2013.09.011

[179] Li X, Shi H, Wang Y, Zhang S, Chu J, Zhang M, et al. Effects of vitamins (nicotinic acid, vitamin B1 and biotin) on phototrophic hydrogen production by *Rhodobacter sphaeroides* ZX-5. International Journal of Hydrogen Energy. 2011;**36**:9620-9625. DOI: 10.1016/j.ijhydene.2011.04.219

[180] Rucker RB, Morris J, Fascetti AJ. Chapter 23-Vitamins. In: Kaneko JJ, Harvey JW, Bruss ML, editors. Clinical Biochemistry of Domestic Animals. 6th ed. Academic Press; 2008. pp. 695-730. DOI: 10.1016/B978-0-12-370491-7.00023-4

[181] Alfenore S, Molina-Jouve C, Guillouet SE, Uribelarrea JL, Goma G, Benbadis L. Improving ethanol production and viability of *Saccharomyces cerevisiae* by a vitamin feeding strategy during fed-batch process. Applied Microbiology and Biotechnology. 2002;**60**:67-72. DOI: 10.1007/s00253-002-1092-7

[182] Zhu MJ, Liang SZ. Effects of biotin on cell growth and astaxanthin accumulation of *Phaffia rhodozyma*. Journal of South China University of Technology (Natural Science). 2003;**31**: 80-84

[183] Mital KM. Biogas Systems: Principles and Applications. New Delhi: New Age International (P) Limited Publishers; 1996

[184] Yadvika S, Sreekrishnan TR, Kohli S, Rana V. Enhancement of biogas production from solid substrates using different techniques-A review. Bioresource Technology. 2004;**95**:1-10. DOI: 10.1016/j.biortech.2004.02.010

[185] Chandra R, Takeuchi H, Hasegawa T. Methane production from lignocellulosic agricultural crop wastes: A review in context to second generation of biofuel production. Renewable and Sustainable Energy Review. 2012;**16**:1462-1476. DOI: 10.1016/j.rser.2011.11.035

[186] Yu HQ, Fang HHP. Acidogenesis of dairy wastewater at various pH levels. Water Science & Technology. 2002;**45**: 201-206

[187] Kim M, Gomec CY, Ahn Y, Speece RE. Hydrolysis and acidogenesis of particulate organic material in mesophilic and thermophilic anaerobic digestion. Journal Environmental Technology. 2003;**24**:1183-1190. DOI: 10.1080/09593330309385659

[188] Dague RR. Application of digestion theory to digester control. Water

Pollution Control Federation. 1968;**40**: 2021-2032

[189] Ward AJ, Hobbs PJ, Holliman PJ, Jones DL. Optimisation of the anaerobic digestion of agricultural resources. Bioresource Technology. 2008;**99**: 7928-4790. DOI: 10.1016/j.biortech.2008.02.044

[190] Deublein D, Steinhauser A. Biogas from Waste and Renewable Resources: An Introduction. Weinheim: Wiley-VCH Verlag GmbH & Co. KGaA; 2008. DOI: 10.1002/9783527621705

[191] Neves L, Oliveira R, Alves MM. Influence of inoculum activity on the bio-methanization of a kitchen waste under different waste/inoculum ratios. Process Biochemistry. 2004;**39**: 2019-2024. DOI: 10.1016/j.procbio.2003.10.002

[192] Prashanth S, Kumar P, Mehrotra I. Anaerobic degradability: Effect of particulate COD. Journal of Environmental Engineering. 2006;**132**: 488-496. DOI: 10.1061/(ASCE)0733-9372

[193] Eiroa M, Costa JC, Alves MM, Kennes C, Veiga MC. Evaluation of the biomethane potential of solid fish waste. Waste Management. 2012;**32**:1347-1352. DOI: 10.1016/j.wasman.2012.03.020

[194] Elbeshbishy E, Nakhla G, Hafez H. Biochemical methane potential (BMP) of food waste and primary sludge: Influence of inoculum pre-incubation and inoculum source. Bioresource Technology. 2012;**110**:18-25. DOI: 10.1016/j.biortech.2012.01.025

[195] Chen TH, Hashimoto AG. Effects of pH and substrate: Inoculum ratio on batch methane fermentation. Bioresource Technology. 1996;**56**:179-186. DOI: 10.1016/0960-8524(96)00016-8

[196] Cantrell KB, Ducey T, Ro KS, Hunt PG. Livestock waste-to-bioenergy generation opportunities. Bioresour Technology. 2008;**99**:7941-7953. DOI: 10.1016/j.biortech.2008.02.061

[197] Owen WF, Stuckey DC, Healy JB, Young LY, McCarty PL. Bioassay for monitoring biochemical methane potential and anaerobic toxicity. Water Research. 1979;**13**:485-492. DOI: 10.1016/0043-1354(79)90043-5

[198] Mao C, Feng Y, Wang X, Ren G. Review on research achievements of biogas from anaerobic digestion. Renewable and Sustainable Energy Review. 2015;**45**:540-555. DOI: 10.1016/j.rser.2015.02.032

[199] Van Ginkel SW, Logan B. Increased biological hydrogen production with reduced organic loading. Water Research. 2005;**39**: 3819-2386. DOI: 10.1016/j.watres.2005.07.021

[200] Lay CH, Wu JH, Hsiao CL, Chang JJ, Chen CC, Lin CY. Biohydrogen production from soluble condensed molasses fermentation using anaerobic fermentation. International Journal of Hydrogen Energy. 2010;**35**: 13445-13451. DOI: 10.1016/j.ijhydene.2009.11.128

[201] Hobson PN, Bousfield S, Summers R. Methane production from agricultural and domestic wastes. London: Applied Science Publishers; 1981. DOI: 10.1007/978-94-009-8102-7

[202] Guwy AJ, Hawkes FR, Wilcox SJ, Hawkes DL. Neural network and on-off control of bicarbonate alkalinity in a fluidized-bed anaerobic digester. Water Research. 1997;**31**:2019-2025. DOI: 10.1016/S0043-1354(97)00016-X

[203] National Research Council. Methane generation from human, animal and agricultural wastes. Washington, DC: The National Academies Press; 1977. DOI: 10.17226/21482

[204] Ren N, Wang A, Cao G, Xu J, Gao L. Bioconversion of lignocellulosic biomass to hydrogen: Potential and challenges. Biotechnology Advances. 2009;**27**:1051-1060. DOI: 10.1016/j.biotechadv.2009.05.007

[205] Lee DJ, Show KY, Su A. Dark fermentation on biohydrogen production: Pure culture. Bioresource Technology. 2011;**102**:8393-8402. DOI: 10.1016/j.biortech.2011.03.041

[206] Ghosh S, Dairkee UK, Chowdhury R, Bhattacharya P. Hydrogen from food processing wastes via photo-fermentation using Purple Non-sulfur Bacteria (PNSB)—A review. Energy Conversion and Management. 2017;**141**:299-314. DOI: 10.1016/j.enconman.2016.09.0

[207] Yokoi H, Mori S, Hirose J, Hayashi S, Takasaki Y. H_2 production from starch by a mixed culture of *Clostridium butyricum* and *Rhodobacter* sp. M-19. Biotechnology Letters. 1998;**20**:895-899. DOI: 10.1023/a:1005327912678

[208] Fang HHP, Zhu H, Zhang T. Phototrophic hydrogen production from glucose by pure and co-cultures of *Clostridium butyricum* and *Rhodobacter sphaeroides*. International Journal of Hydrogen Energy. 2006;**31**:2223-2230. DOI: 10.1016/j.ijhydene.2006.03.005

[209] Asada Y, Tokumoto M, Aihara Y, Oku M, Ishimi K, Wakayama T, et al. Hydrogen production by co-cultures of Lactobacillus and a photosynthetic bacterium, *Rhodobacter sphaeroides* RV. International Journal of Hydrogen Energy. 2006;**31**:1509-1513. DOI: 10.1016/j.ijhydene.2006.06.017

[210] Laurinavichene TV, Laurinavichius KS, Tsygankov AA. Integration of purple non-sulfur bacteria into the starch-hydrolyzing consortium. International Journal of Hydrogen Energy. 2014;**39**:7713-7720. DOI: 10.1016/j.ijhydene.2014.03.088

[211] Ren N, Guo W, Liu B, Cao G, Ding J. Biological hydrogen production by dark fermentation: Challenges and prospects towards scaled-up production. Current Opinion in Biotechnology. 2011;**22**:365-370. DOI: 10.1016/j.copbio.2011.04.022

[212] Madsen M, Holm-Nielsen JB, Esbensen KH. Monitoring of anaerobic digestion processes: A review perspective. Renewable and Sustainable Energy Reviews. 2011;**15**:3141-3155. DOI: 10.1016/j.rser.2011.04.026

[213] Abbasi T, Tauseef SM, Abbasi SA. Anaerobic digestion for global warming control and energy generation-An overview. Renewable and Sustainable Energy Reviews. 2012;**16**:3228-3242. DOI: 10.1016/j.rser.2012.02.046

[214] Liu X, Bayard R, Benbelkacem H, Buffière P, Gourdon R. Evaluation of the correlations between biodegradability of lignocellulosic feedstocks in anaerobic digestion process and their biochemical characteristics. Biomass and Bioenergy. 2015;**81**:534-543. DOI: 10.1016/j.biombioe.2015.06.021

[215] Raposo F, Fernández-Cegrí V, De la Rubia MA, Borja R, Béline F, Cavinato C, et al. Biochemical methane potential (BMP) of solid organic substrates: Evaluation of anaerobic biodegradability using data from an international interlaboratory study. Journal of Chemical Technology & Biotechnology. 2011;**86**:1088-1098. DOI: 10.1002/jctb.2622

[216] Nielfa A, Cano R, Fdz-Polanco M. Theoretical methane production generated by the co-digestion of organic fraction municipal solid waste and biological sludge. Biotechnology Reports. 2015;**5**:14-21. DOI: 10.1016/j.btre.2014.10.005

[217] Li Y, Zhu J, Wan C, Park SY. Solid-state anaerobic digestion of corn stover for biogas production. Transactions of the ASABE. 2011;**54**: 1415-1421. 10.13031/2013.39010

[218] Brown D, Shi J, Li Y. Comparison of solid-state to liquid anaerobic digestion of lignocellulosic feedstocks for biogas production. Bioresource Technology. 2012;**124**:379-386. DOI: 10.1016/j.biortech.2012.08.051

[219] Sawatdeenarunat C, Surendra KC, Takara D, Oechsner H, Khanal SK. Anaerobic digestion of lignocellulosic biomass: Challenges and opportunities. Bioresource Technology. 2015;**178**: 178-186. DOI: 10.1016/j.biortech. 2014.09.103

[220] Liu ZL, Saha BC, Slininger PJ. Lignocellulosic biomass conversion to ethanol by Saccharomyces. In: Wall J, Harwood C, Demain A, editors. Bioenergy. Washington, DC: ASM Press; 2008. pp. 17-36. DOI: 10.1128/ 9781555815547.ch2

Chapter 7

Selection of Optimal Localization for a Biomass Energy Plant that Uses Residual Biomass as a Raw Material in the Araucanía Region of Chile

Celián Román-Figueroa, Sebastián Herrera and Manuel Paneque

Abstract

Residual biomass is used for energy purposes, such as producing heat and electricity. The potential raw materials are wheat, oat and barley straw, corn stover from the agricultural industry, and wood chips from forest residuals and the wood industry. In Chile about 10 MM t year^{-1} residues are produced; however, most of these are burned, increasing the greenhouse gas emission. This study evaluated the energy production from the residual biomass in the Araucanía Region of Chile. The optimal location for installing an energy generation plant was determined, according to various conditions, such as the distance from the villages, energy demands, industry location, etc. A multi-criteria evaluation and analytic hierarchy process were performed for determining the optimal location, and a proposal for its installation was created. A total of 19 districts were selected as suitable location that met the requirements for the three scenarios.

Keywords: biomass, gasification, combustion, wheat straw, bioenergy, energy production

1. Introduction

A close relationship exists between an increase in energy consumption and the economic growth for a particular country. In South America, it is estimated that a 1% increase in energy consumption translates to a 0.42% increase in economic growth [1]. In countries, such as Chile, where 59.3% of the primary energy matrix is imported, with 90.2% of this from fossil fuels [2], uncertainty and insecurity arise in the energy supplies and dependence on foreign markets, obliging the assumption of risk in the face of possible fluctuations [1].

Alternative sources of energy that are renewable and have a reduced environmental impact are required to reduce the use of nonrenewable energy sources, such as fossil fuels [3], which have negative effects, such as climate change, forest destruction, and the extinction of species [3, 4].

 IntechOpen

Renewable energy offers environmental benefits and increases the standard of living for various populations, diversifies the energy matrix, improves the infrastructure, promotes technology transfer, and provides other positive effects [4, 5]. Biofuels are nonconventional renewable energy sources (NCRECs) that may replace fossil fuels, lowering the dependence on international markets and the atmospheric emissions of greenhouse gases (EGGs) [6].

Chile promised, in 2010, to reduce emissions by 20% below the 2020 projection [7]; however, according to current trends, an increase of 360% is projected in carbon dioxide (CO_2) emissions in electricity generation and transportation. The aforementioned items currently represent 0.3% of the total emissions globally [8].

Firewood and biofuels are the second source of energy for Chile. They are entirely produced within the country, representing 28.9% of the primary energy matrix in 2013 [9]. Here, 36% of the national population is concentrated between the O'Higgins and Aysén Regions, with 74% of these homes consuming firewood or its derivatives [10] for heating or cooking systems. Biofuel is used for self-produced electricity [11].

The residual biomass from agricultural activities has an average caloric power of 17,500 kJ kg^{-1} [12] and is underused in Chile [2, 13]. Cereal production residue is concentrated in the central south area of the country, particularly in the regions of Araucanía (29.3%) and Libertador Bernardo O'Higgins (19.8%) as the main cereal residue producers [14].

Wheat is the main cereal produced in Chile, representing 32.9% of the planted agricultural surface during the 2016/2017 agricultural season [15]. The Araucanía Region is the main wheat production region in the country. During the 2016/2017 agricultural season, 42.0% of the total surface area of planted wheat was concentrated here, yielding a production of approximately 597,835 tons [16].

Román-Figueroa et al. [2] determined that, in the Araucanía Region, 50% of the production of wheat residuals was concentrated in 23 (of 299) census districts, while 10 of these districts produced 27.8% of the residues. The majority was produced in the central valley of the region, specifically in the province of Malleco, which has a regional coverage of 60,800 ha [2]. Currently, the agricultural residue is burnt [17, 18], which causes environmental problems owing to the emission of EGGs, as well as public health problems owing to particulate matter emissions [17, 19].

Electricity production from agricultural residue biomass has been widely studied and recommended, owing to the low production costs, high conversion efficiency, and environmental benefits because it is carbon neutral [20, 21]. Singh [21] determined, in the Punjab, India region, production of between 2375 and 2937 MW_{el} was possible depending on the efficiency of the conversion plant, with more than 22,000 million tons of residue. In the Araucanía Region, a 5.0 MW_{th} plant and 27,000 tons of residue, between 3.17 MW_{el} and 4.89 MW_{el}, can be produced using fluidized bed combustion technology with a generation turbine (C/ST) and gasification of the fluidized bed followed by a combined cycle of gas and vapor (G/CC), respectively [2].

Various studies have determined the optimal location of a biomass-based energy production plant using geographic information systems (GIS) [21–25]. With GIS, evaluation using different attributes and maps to determine the optimal energy production plant location is feasible, [23, 25]. A multi-criteria analysis (MCA) evaluates, using different criteria or factors, a group of opposed real alternatives, considering different development visions and objectives [25]. Therefore, an evaluation considering economic, social, and environmental criteria is possible, optimizing the decision-making process [23].

In Chile, Villamar et al. [23] evaluated the possibility of installing an anaerobic co-digestion plant using discarded agribusiness materials (animal dung and agricultural residues) in the Biobío Region. Using a hierarchical analysis process, they considered factors that were social (distances to residential areas and roads) and economic (residue production, distance to residue production sites, proximity between residue production sites, and closeness of the production plant to roads) [23]. This is the only evaluation of the installation of a residual biomass-based bioenergy plant in the country.

The objective of this study was to determine the optimal location for an energy production plant, which was based on wheat residue in the Araucanía Region of Chile. Three different types of demands were considered: current, potential, and social demands. Three scenarios were used to determine the location of the energy production plant based on wheat residue biomass.

2. Materials and methods

2.1 Area of study

The area of study was the Araucanía Region, located between 37°35′ and 39°37′ southern latitude and from 70°50′ western longitude to the Pacific Ocean, an area of 31,842 km^2 (**Figure 1**). The study was realized at a district level

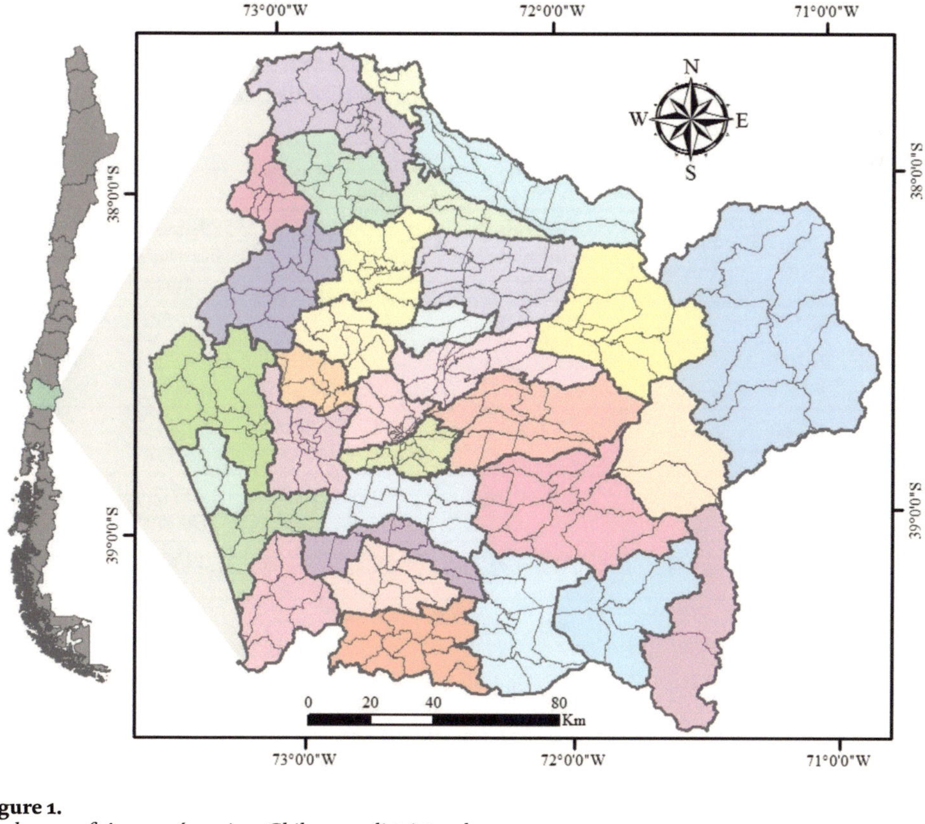

Figure 1.
Study area of Araucanía region, Chile, on a district scale.

according to the census boundaries determined by the National Institute of Statistics [26].

2.2 Areas with an energy demand

The areas were determined with priority assigned to the energy supply, according to three types of demand: current (or high demand), potential, and social. The current demand refers to the sectors with a high consumption. The potential demand refers to the non-covered demand, such as the districts without electricity. The social demand refers to areas where heating or electrical energy should be locally supplied to homes, schools, or treatment centers.

Different criteria were employed to determine the demand in each case. The current demand considers consumers in the industrial [27] and residential sectors [28]. For the potential demand, the demand for electrical services that are not yet provided is considered [29]; and in the social demand, the demand for energy (thermal and electrical) for services (education and healthcare) and rural homes is considered.

2.2.1 Standardization of criteria for determining the areas with an energy demand

Each criterion was standardized with values between 1 and 7, with 1 representing the worst condition and 7 representing the best condition for each criterion.

For the current demand, the distance to the industries, industries quantity per district [30], and the residential energy consumption were considered [30, 31]. All criteria were discretized with values between 1 and 7 (**Table 1**). For the potential demand, those districts with an electrical coverage of less than 40% were considered, according to the 2006 CASEN [28]. These were discretized with values between 1 and 7. However, only odd values were used owing to the quantity of

Value	Current			Potential	Social		
	Industries (m)	Industries (number)	Energy consumption (MW)	Coverage (%)	Villages (m)	Primary healthcare (m)	Rural educational (m)
7	0–21,300	7–8	2000–12,000	0–10	0–9531	0–4572	0–4019
6	21,300–42,600	6	1600–2000	—	9531–19,062	4572–9145	4019–8038
5	42,600–63,900	5	1300–1600	10–20	19,062–28,593	9145–13,717	8038–12,057
4	63,900–85,201	4	1000–1300	—	28,593–38,124	13,717–18,289	12,057–16,076
3	85,201–106,501	3	700–1000	20–30	38,124–47,655	18,289–22,861	16,076–20,095
2	106,501–127,801	2	400–700	—	47,655–57,186	22,861–27,434	20,095–24,114
1	127,801–149,101	1	0–400	30–40	57,186–66,717	27,434–32,006	24,114–28,133

Table 1.
Discretization of the values to determine the supply of energy in the Araucanía region, according to the different demand criteria.

Current demand		Potential demand		Social demand	
Criterion	**Weighting**	**Criterion**	**Weighting**	**Criterion**	**Weighting**
Distance to the industries	0.25	Electrical coverage	1.0	Primary healthcare	0.33
Industry quantity	0.25			Rural educational	0.33
Energy consumption	0.5			Villages	0.33

Table 2.
Consideration of the criteria analyses in the MCE to determine the locations with the highest energy demand (high, potential, and social).

ranges available (**Table 1**). For the social demand, the distance to rural settlements, which corresponded to the villages, as well as the distances to primary healthcare centers and rural educational establishments were considered [32]. These were discretized with values between 1 and 7, with 7 representing the shortest distance (**Table 1**).

Based on the type of demand and the multi-criteria evaluation (MCE), weights or considerations were assigned to each criterion (**Table 2**). The priority sites were determined, considering each type of demand (current, potential, and social). Finally, the set of demands was evaluated, where each demand was given a consideration value of 0.33. Thereafter, the energy demand information was used to determine the location of an energy production plant based on biomass in the Araucanía Region.

2.3 Determination of the priority sites for installing an energy production plant based on biomass

Determining the location of an energy production plant based on biomass was carried out according to a decision rule, where the MCE was used with an analytic hierarchy process (AHP). The use of these combined techniques is a common methodology employed to determine the installation location of energy production plants [29, 33].

2.3.1 Definition of criteria

Criteria selection is required for making decisions. Possible alternatives must be quantified and must contribute to the decision-making process [33–35]. These criteria may be factors or limiters. A factor is a criterion that negatively or positively affects the possible location for the energy production plant, while a limiter (or exclusion criterion) restricts the location potential, excluding possible installation [33, 34].

The factors employed in this evaluation were as follows:

a. Availability of biomass: The districts with a greater production of wheat residue biomass according to the estimates by Román-Figueroa et al. [2] were considered the most appropriate; however, this was not established as an exclusion criterion.

b. Energy consumption priority areas: These were established according to the energy demand scenarios previously calculated: high demand, social demand,

and potential demand. The selected sectors with a high demand were considered a priority and more suitable.

c. Land use/coverage: The installation of an energy production plant must comply with the legal requirements and land usage. Competition with other current uses of the land must be avoided, allowing current activities to continue. The areas considered feasible for the installation of a plant were those without vegetation, meadows, or scrub and were industrial use lands. Urban areas and lands of agricultural use were considered unsuitable because other activities are carried out there that are relevant to society [33, 34].

d. Roads: Connectivity is essential for the supply of raw materials to the plant; thus, the transport of raw materials should be considered [33, 36]. Therefore, a maximum distance of 1.0 km between the energy plant and the road was considered a priority. Moreover, the type of road material affects accessibility, and thus, this was considered a factor.

e. Electrical network: Wheat residue was considered the raw material for electricity production [2]. The distance from the energy production plant to the electrical network is relevant to the electricity generated in the electrical supply network. Nonconventional renewable energy (NCRE), such as a generation less than 9.0 MW_{el}, can directly connect to the Chilean distribution network [37]. This would provide an economic benefit to plant installation. A maximum distance of 3.0 km between the energy plant and the electrical network was considered a priority.

Each of these factors was discretized with values between 1 and 7, with 7 as the most suitable (**Table 3**).

Conversely, the following limiters were considered:

a. Restricted areas: This represents zones where installation is not viable for environmental, legal, or other reasons [33, 34]. These areas include those covered by the protected forest areas system (SNASPE), protected private areas (PPAs), bodies of water, cities, and the current road network. A value of 0 was assigned to the limiter areas, and 1 was assigned to those without restrictions.

Value	Residual biomass (t $year^{-1}$)	Land use/coverage	Road types	Electrical network (m)
7	23,217–27,086	Grassland and scrubland	Asphalt	0–1000
6	19,347–23,217	—	—	1000–2000
5	15,478–19,347	Industrial and urban area	Gravel	2000–3000
4	11,608–15,478	—	—	3000–12,000
3	7739–11,608	Primary and secondary forest	Dirt	12,000–21,000
2	3869–7730	—	—	—
1	0–3869	Agricultural land and plantations	—	—

Table 3.
Discretization of the values for the factors used to determine the optimal location of an energy production plant based on the wheat biomass in the Araucanía region.

b. Availability of biomass: Districts where the minimum quantity of wheat residue is required for the production of 1 MW_{el} were considered as feasible, such as the districts with a residue production less than 8534.8 t $year^{-1}$, for a plant with fluidized bed combustion and a generating turbine (C/ST), and 5536.1 t $year^{-1}$, for a plant with fluidized bed gasifiers followed by a combined cycle of gas and steam (G/CC), based on the study carried out by Román-Figueroa et al. [2]. A value of 0 was assigned to the districts with residue production less than the minimum required to produce 1 MW_{el}, while other districts with higher production were assigned a value of 1. Depending on the situation, the availability of biomass was used as a factor, with the production of biomass being classified with values from 1 to 7 (**Table 3**). C/ST and G/CC were considered independently as technologies for energy production.

2.3.2 Proposals for energy production plant placements according to the energy demand scenarios

An AHP was used to determine the best location for an energy production plant using wheat biomass as a raw material. The main criterion for the plant location was the biomass supply, by which other major criteria were determined. The analysis was carried out considering the energy demand scenarios (current, potential, and social) as part of the analysis.

In the AHP, an importance value was assigned to each criterion according to its objective. The value scale ranged from 1 to 9, where 1 was considered to be equally important for two criteria (a and b) and 9 was considered to be more important for one criterion than another (a over b) [38, 39]. Therefore, when the relative importance of a over b was established, a value of 1/9 was automatically established [39].

a. Social scenario: The objective was to satisfy rural sector energy needs, supplying services to healthcare, educational, and rural resident sectors. This scenario was based on social and potential demand, and therefore, industrial and residential (that already have supply) sectors were not considered in this scenario. The availability of biomass, use of land, roads, and electrical supply network were considered as factors. Each of these factors was considered with a different relative weight (**Table 4**).

b. Feasibility scenario: The objective was to install an energy production plant based on the biomass, and thus, the residual biomass was considered as a limiter, 8534.8 t $year^{-1}$ for C/ST and 5536.1 t $year^{-1}$ for G/CC [2]. The areas with an energy demand had less relevance than other criteria because the availability of the raw material is most relevant (**Table 4**).

c. Demand scenario: The objective was to satisfy large consumer demand; therefore, the high-demand scenario was used, where priority was assigned to industrial and residential sectors that currently receive electricity. This was not considered relevant for rural sectors that do not receive. The electrical network had a greater relevance in this scenario because of the goal to satisfy the current energy demand (**Table 4**).

Criterion	**Social scenario**					
	Biomass	**Social demand**	**Potential demand**	**Land use**	**Roads**	**Electrical network**
Biomass	1	3	3	5	5	5
Social demand	1/3	1	1	3	3	3
Potential demand	1/3	1	1	3	3	3
Land use	1/5	1/3	1/3	1	1	1
Road	1/5	1/3	1/3	1	1	1
Electrical network	1/5	1/3	1/3	1	1	1
	Feasibility scenario					
	Priority areas	Land use	Roads	Electrical network		
Priority areas	1	1/7	1/7	1/7		
Land use	7	1	1	1		
Road	7	1	1	1		
Electrical network	7	1	1	1		
	Demand scenario					
	Biomass	Current demand	Land use	Roads	Electrical network	
Biomass	1	3	7	7	7	
Current demand	1/3	1	7	7	7	
Land use	1/7	1/7	1	1	1	
Road	1/7	1/7	1	1	1	
Electrical network	1/7	1/7	1	1	1	

Table 4.
Valorization of the different criteria used to determine the optimal location of a production plant based on the biomass according to the social, feasibility, and demand scenarios.

3. Results and discussion

3.1 Areas with energy demand

a. High demand: This scenario was created based on the number of industries, distance to the industry, and current energy consumption existing in the region. The industries in the Araucanía Region are concentrated in the municipalities of Angol and Temuco. This is because the municipalities are the capitals of the provinces of Malleco and Cautín, respectively. They are also the cities with a large labor and economic importance, along with Villarica, in the Araucanía Region [40]. Conversely, the current energy consumption in the region generally occurs in the municipalities of Temuco, Ercilla, Cunco,

Perquenco, and Freire and to a lesser degree in Angol, Villarrica, and Pucón. These municipalities concentrate the populations; besides, they are located in Ruta 5 road or they are the main touristic centers in the Araucanía Region [40]. Because of the greater consideration placed on the residential electrical consumption (0.5) than that of the quantities of and distances to the industries (0.25 for each one; **Table 2**), the high-demand scenario concentrated on the same municipalities (**Figure 2a**).

b. Potential demand: This scenario was created based on the existing rural electricity required in the Araucanía Region, which is one of the regions with the least amount of rural electricity in Chile [41]. Therefore, the districts with electrical coverage less than 40% were given priority. These districts are principally located in the mountain or costal area of the region and the extreme north and south. The total surface area of the districts with these characteristics was 447,954 ha (**Figure 2b**).

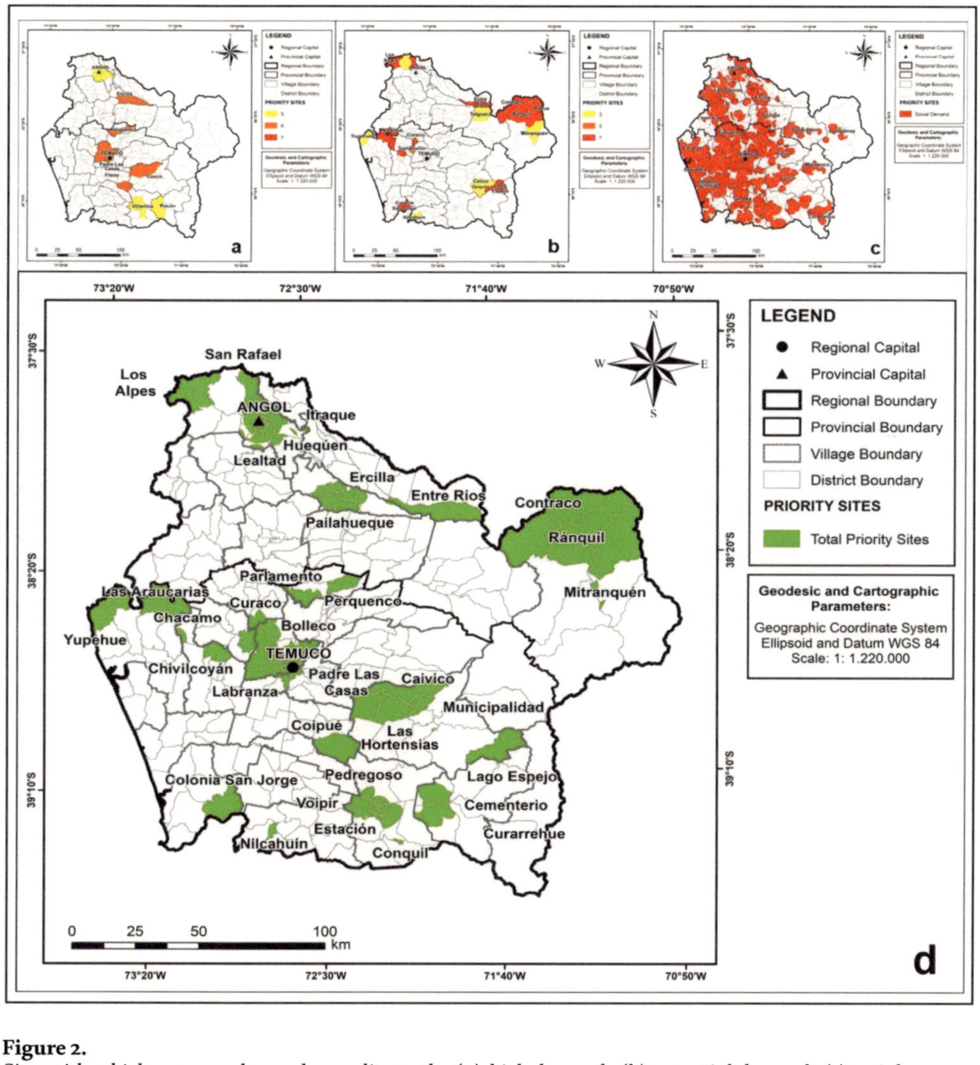

Figure 2.
Sites with a higher energy demand according to the (a) high demand, (b) potential demand, (c) social demand scenarios, and (d) priority energy demand sites according to the three demand scenarios in the Araucanía region.

c. Social demand: This scenario was based on rural establishments (educational and healthcare facilities and villages), prioritizing the energy supply to these locations. All priority sites under this scenario were located following the central axis with the most populated municipalities, Angol, Victoria, and Temuco [40], distributing to west and southwest of Temuco, covering rural coast municipalities, Carahue, Teodoro Schmidt, Saavedra, and Nueva Toltén (**Figure 2c**).

Information for each of the scenarios was cross-matched linearly to determine the priority energy demand sites. The priority sites occupied an irregular distribution in the territory (**Figure 2d**) and covered approximately 548,134 ha (17.2% of the regional surface).

3.2 Priority sites for installing an energy production plant based on biomass in the Araucanía region

The availability of the wheat biomass residue, considered a biomass production of 8534.8 t year^{-1} for C/ST and 5536.1 t year^{-1} for G/CC in the Araucanía Region, was concentrated in the central valley, in the south of the Malleco Province and in the north of the Cautín Province (**Figure 3a**) [2]. The usable land for installing an energy production plant is mainly located in the peripheral areas of the region, to the south and in the Andes Cordillera. The central area has a greater level of agricultural and industrial development (**Figure 3b**); however, the distribution of

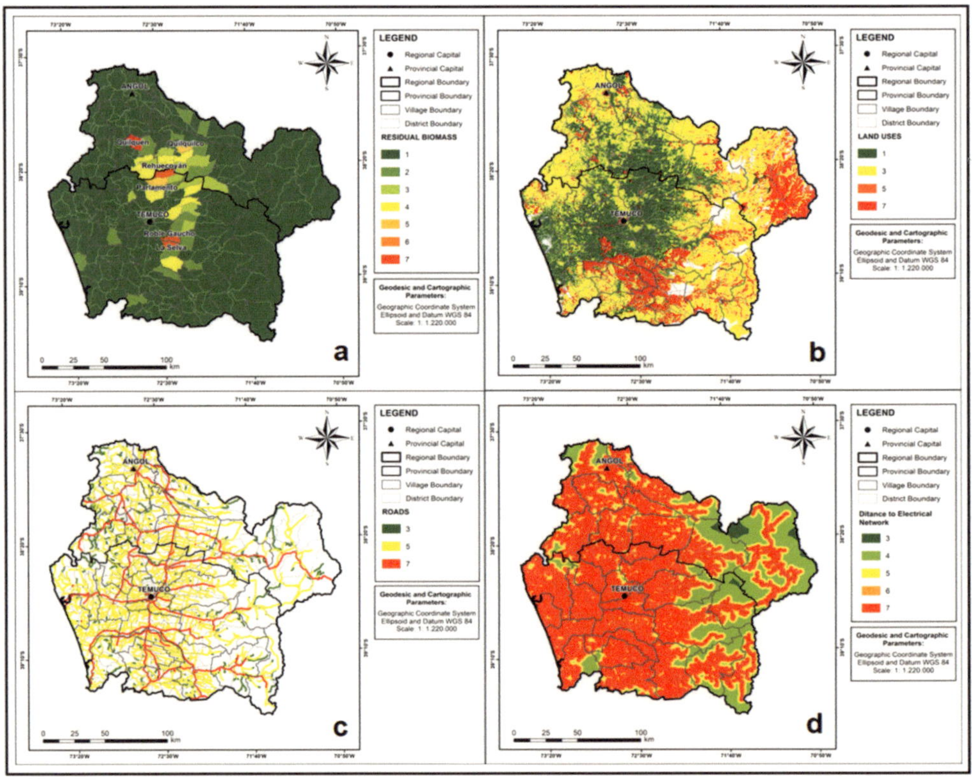

Figure 3.
Prioritization of the criteria established for the installation of an energy production plant based on the wheat residue biomass in the Araucanía region: (a) biomass production on a district-wide basis, (b) land use, (c) road network, and (d) distance to the electricity distribution network.

the development could be modified by reducing the use of scrubland meadows; the locations with a higher priority for an energy plant may be hindered by agricultural use [42].

The road network is a central paved axis, which corresponds to Route 5 that runs from the north to the south of the region. There are a series of secondary paved roads that run perpendicular to this route. Generally, paved roads are scarce, while gravel roads represent the densest network of roads in the region and are distributed toward the coastal area. Finally, dirt roads are found in the peripheral areas and are scarce (**Figure 3c**). The distance to the electrical distribution network displayed a similar pattern to that of the road network. A concentration of smaller distances to the network in the central and west areas was observed, while places with larger distance were found in the northeast region. Distances greater than 21,000 m were recorded (**Figure 3d**).

3.2.1 Determination of the priority placement areas

Standardized matrixes and priority vectors were established (or standardized vectors) for each of the priority placement scenarios (social, feasibility, and demand) for each energy production plant based on the wheat biomass (**Table 5**).

An evaluation of the consistency in the priority evaluation matrices is relevant in AHP to avoid inconsistencies in the final matrix [33, 38, 43, 44]. Thus, a consistency reason (CR) was used. A consistency index (CI) and a randomness index (RI) were used to calculate the CR [33, 38, 44]. The CRs in the social and demand scenario matrices were 0.019 (CI: 0.023 and RI: 1.24) and 0.049 (CI: 0.056 and RI: 1.12), respectively. This showed that the matrices were consistent, since the inconsistencies of the chosen weights were less than 10% [38, 44]. For the feasibility scenario matrix, the column values were the same for all criteria, showing that the weights (values) used were consistent; therefore, a consistency analysis does not need to be performed [44, 45]. The CR would be equal to 0. Mu and Pereyra-Rojas [44] recommended performing a consistency analysis whenever the criteria were more than two. The CR for the feasibility scenario was 0 because the CI values obtained from the priority vector was 0.

Value of priority vectors were considered as the final relative weight to each one of the variables in this analysis, for determining the optimal locations according to the three priority site scenarios [33, 43]. Here, under the social site scenario, the biomass availability was the most relevant (41.2%), while under the social demand and potential scenarios, the biomass availability was secondary (18.7% relevance for each). The same pattern was found in the demand scenario, where the availability was most relevant (49.6%), followed by the high-energy demand (33.0%). Finally, in the feasibility scenario, where biomass was considered a limiter, the land use, road network, and distance to the electrical distribution network were considered the factors with the greatest weight (31.8%).

Using the three scenarios for determining the best location for an energy production plant based on the wheat residue biomass, a similar behavior was observed in the priority establishment sites (**Figure 4**). In the social scenario, the optimal locations were dispersed in the central valley of the region, with a greater quantity in the south of the province of Malleco and the north of the province of Cautín, covering a surface of 226,414 ha (**Figure 4a**). In the feasibility scenario, the availability of the residue biomass was considered as a limiter; therefore, the best sites were clustered in areas similar to the distribution of the residues, which was concentrated in the central area of the region. For a G/CC plant with a minimum biomass requirement of 5536.1 t year^{-1}, the available surface was 54,795 ha. However, for a C/ST plant, with a minimum biomass requirement of 8534.8 t year^{-1}, the

Criterion	**Social scenario**						
	Biomass	**Social demand**	**Potential demand**	**Land use**	**Roads**	**Electrical network**	**PVE**
Biomass	0.4412	0.5000	0.5000	0.3571	0.3571	0.3571	0.4188
Social demand	0.1471	0.1667	0.1667	0.2143	0.2143	0.2143	0.1872
Potential demand	0.1471	0.1667	0.1667	0.2143	0.2143	0.2143	0.1872
Land use	0.0882	0.0556	0.0556	0.0714	0.0714	0.0714	0.0689
Road	0.0882	0.0556	0.0556	0.0714	0.0714	0.0714	0.0689
Electrical network	0.0882	0.0556	0.0556	0.0714	0.0714	0.0714	0.0689
	Feasibility scenario						
	Priority areas	**Land use**	**Roads**	**Electrical network**			**PVE**
Priority areas	0.0455	0.0455	0.0455	0.0455			0.0455
Land use	0.3182	0.3182	0.3182	0.3182			0.3182
Road	0.3182	0.3182	0.3182	0.3182			0.3182
Electrical network	0.3182	0.3182	0.3182	0.3182			0.3182
	Demand scenario						
	Biomass	**Current demand**	**Land use**	**Roads**	**Electrical network**		**PVE**
Biomass	0.5676	0.6774	0.4118	0.4118	0.4118		0.4961
Current demand	0.1892	0.2258	0.4118	0.4118	0.4118		0.3301
Land use	0.0811	0.0323	0.0588	0.0588	0.0588		0.0580
Road	0.0811	0.0323	0.0588	0.0588	0.0588		0.0580
Electrical network	0.0811	0.0323	0.0588	0.0588	0.0588		0.0580

Table 5.
Standardized matrices and priority vectors used to determine the optimal location of an energy production plant in the Araucanía region, according to the social, feasibility, and demand scenarios.

available surface was 41,949 ha (**Figure 4b**). In the high-demand scenario, a similar distribution to that of the feasibility scenario was observed, with a concentration of the optimal areas in the center of the region. The available space was 183,235 ha (**Figure 4c**).

Finally, 19 districts that met the requirements for the three scenarios were found. Thus, the installation of the plant may be considered an integrated solution for different types of demand (social, feasibility, and high demand). These districts were Chufquén, Coipué, Colonia Lautaro, Dollinco, General López, Huichahue, La Colmena, La Selva, Manzanaco, Parlamento, Perquenco, Quilquén, Quilquilco, Quino, Quintrilpe, Rehuecoyán, Roble Gaucho, Santa Ana, and Tricauco (**Figure 4d**).

In this study, the possibility of producing electrical energy using wheat residue biomass was evaluated based on two different combined cycle technologies

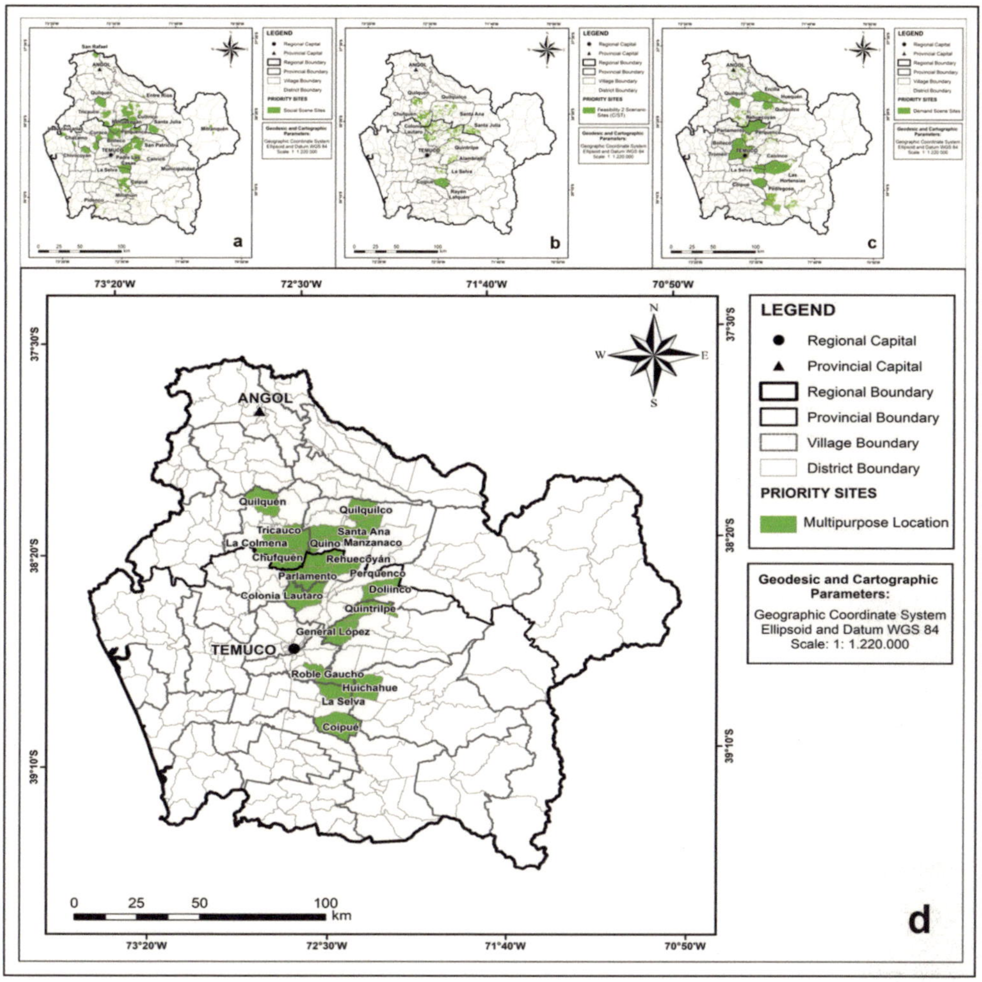

Figure 4.
Priority sites for energy plant installation an energy production plant based on the wheat biomass in the Araucanía region obtained by AHP under different scenarios, (a) social scenario; (b) feasibility scenario for C/ST plant with 8534.8 t year^{-1} biomass as minimum requirement; (c) current demand scenario; and (d) multipurpose priority sites based in three scenarios for installation an energy production plant based on the wheat biomass in the Araucanía region.

(C/ST and G/CC), where the biomass was submitted to a thermic process and then generated electricity [2]. These technologies must be economically evaluated to determine their viability in the Araucanía Region. In addition, there are other potential uses for residual biomass. Recently, Azócar et al. [46] evaluated the production of pellets from wheat residue biomass using a roasting process that optimized energy production from this raw material. Therefore, there are multiple alternatives for creating energy from this raw material that should be evaluated to avoid burning these residues and the environmental consequences from these actions [17].

4. Conclusions

The multi-criteria analysis determined the sites where energy demand was present based on various scenarios. The energy demand contained a distribution in different sectors of the region. The current demand (high demand) was concentrated in the main economic and residential centers of the region, and because the residential demand was prioritized over the industrial demand, the municipalities of Carahue, Cunco, and Temuco had the highest demand. The potential demand was found in the rural zones, where there is a minimal electrical service. These areas are mainly located in the mountain and/or coastal zones, which are far from the city of Temuco, the regional capital. The social demand showed greater distribution in the region, because there are dispersed rural settlements that require energy.

The optimal location for an energy production plant was determined using relevant information, such as the availability of the biomass, road network, etc. The biomass availability was the criterion with the greatest influence on a placement decision in the three scenarios (social, feasibility, and demand). Therefore, in the three scenarios, the optimal locations had a distribution similar to the availability of the biomass. The districts were Chufquén, Coipué, Colonia Lautaro, Dollinco, General López, Huichahue, La Colmena, La Selva, Manzanaco, Parlamento, Perquenco, Quilquén, Quilquilco, Quino, Quintrilpe, Rehuecoyán, Roble Gaucho, Santa Ana, and Tricauco, where the energy-producing plant, using wheat residue biomass, had greater viability because it meets the demand for the three scenarios under evaluation.

Author details

Celián Román-Figueroa[1,2], Sebastián Herrera[1] and Manuel Paneque[3*]

1 Agroenergía Ingeniería Genética S.A. Inc., Santiago, Chile

2 Doctoral Program in Sciences of Natural Resources, Universidad de La Frontera, Temuco, Chile

3 Department of Environmental Sciences and Natural Resources, Faculty of Agricultural Sciences, University of Chile, Santiago, Chile

*Address all correspondence to: mpaneque@uchile.cl

References

[1] Apergis N, Payne JE. Energy consumption and growth in South America: Evidence from a panel error correction model. Energy Economics. 2010;**32**:1421-1426. DOI: 10.1016/j.eneco.2010.04.006

[2] Román-Figueroa C, Montenegro N, Paneque M. Bioenergy potential from crop residue biomass in Araucanía region of Chile. Renewable Energy. 2017;**102**:170-177. DOI: 10.1016/j.renene.2016.10.013

[3] Kahia M, Aïssa MSB, Charfeddine L. Impact of renewable and non-renewable energy consumption on economic growth: New evidence from the MENA net oil exporting countries (NOECs). Energy. 2016;**116**:102-115. DOI: 10.1016/j.energy.2016.07.126

[4] Tugcu CT, Tiwari AK. Does renewable and/or non-renewable energy consumption matter for total factor productivity (TFP) growth? Evidence from the BRICS. Renewable and Sustainable Energy Reviews. 2016;**65**:610-616. DOI: 10.1016/j.rser.2016.07.01

[5] Domac J, Richards K, Risovic S. Socio-economic drivers in implementing bioenergy projects. Biomass and Bioenergy. 2005;**28**:97-106. DOI: 10.1016/j.biombioe.2004.08.002

[6] Román-Figueroa C, Paneque M. Ethics and biofuel production in Chile. Journal of Agricultural and Environmental Ethics. 2015;**28**:293-312. DOI: 10.1007/s10806-015-9535-1

[7] Ministerio del Medio Ambiente (MMA), Informe del estado del medio ambiente; 2012. 511 p

[8] O'Ryan R, Díaz M, Clerc J. Consumo de energía y emisiones de GEI en Chile 2007-2030 y opciones de mitigación. Faculty of Physics and Mathematical Sciences, University of Chile; 2009. 97 p

[9] Ministerio de Energía (MINERGIA), Política de uso de la leña y sus derivados para calefacción. Santiago, Chile; 2015. 110 p

[10] Ministerio de Energía (MINERGIA), Balance nacional de energía 2013. 2014. Available on: http://www.minenergia.cl/documentos/balance-energetico.html [Accessed: 10-09-2018]

[11] Paneque M, Román-Figueroa C, Vásquez-Panizza R, Arriaza J, Morales D, Zulantay M. Bioenergía en Chile. University of Chile and Food and Agriculture Organization of the United Nations (FAO). Santiago, Chile; 2011. 126 p

[12] Scarlat N, Matinovi M, Dalleman J-F. Assessment of the availability of agricultural crop residues in the European Union: Potential and limitations for bioenergy use. Waste Management. 2010;**30**(10):1889-1897. DOI: 10.1016/j.wasman.2010.04.016

[13] Rodríguez-Monroy C, Mármol-Acitores G, Nilsson-Cifuentes G. Electricity generation in Chile using non-conventional renewable energy sources – A focus on biomass. Renewable and Sustainable Energy Reviews. 2018;**81**:937-945. DOI: 10.1016/j.rser.2017.08.059

[14] Gatica L, Alonso M. Informe Técnico. Disponibilidad y potencial energético de los principales residuos agrícolas de las zonas centro y centro sur de Chile, BIOCOMSA. Faculty of Agricultural Sciences, University of Chile. Santiago, Chile; 2013. 76 p

[15] Oficina de Estudios y Políticas Agrarias (ODEPA). Información nacional de superficie sembrada, producción y rendimientos anuales. 2017. Available on: http://www.odepa.cl/estadisticas/productivas/ [Accessed: 06-09-2018]

[16] Oficina de Estudios y Políticas Agrarias (ODEPA). Información

regional de superficie sembrada, producción y rendimientos anuales. 2017. Available on: http://www.odepa.cl/estadisticas/productivas/ [Accessed: 06-09-2018]

[17] Corporación Nacional Forestal (CONAF), Caracterización de los cultivos y residuos de vegetales derivados de la cosecha de productos forestales y agrícolas en Chile. Santiago, Chile; 2008. 92 p

[18] Damon PM, Bowden B, Rose T, Rengel Z. Crop residue contributions to phosphorus pools in agricultural soils: A review. Soil Biology and Biochemistry. 2014;**74**:127-137. DOI: 10.1016/j.soilbio.2014.03.003

[19] Kavouras IG, Koutrakis P, Cereceda-Balic F, Oyola P. Source apportionment of PM_{10} and $PM_{2.5}$ in five Chilean cities using factor analysis. Journal of the Air and Waste Management Association. 2001;**51**(3):451-464. DOI: 10.1080/10473289.2001.10464273

[20] McKendry P. Energy production from biomass (part 1): Overview of biomass. Bioresource Technology. 2002;**83**:37-46. DOI: 10.1016/S0960-8524(01)00118-3

[21] Singh J. Overview of electric power potential of surplus agricultural biomass from economic, social, environmental and technical perspective e a case study of Punjab. Renewable and Sustainable Energy Reviews. 2015;**42**:286-297. DOI: 10.1016/j.rser.2014.10.015

[22] Perpiñá C, Alfonso D, Pérez-Navarro A, Peñalvo E, Vargas C, Cárdenas R. Methodology based on geographic information systems for biomass logistics and transport optimisation. Renewable Energy. 2009;**34**:555-565. DOI: 10.1016/j.renene.2008.05.047

[23] Villamar CA, Rivera D, Aguayo M. Anaerobic co-digestion plants for the revaluation of agricultural waste: Sustainable location sites from a GIS analysis. Waste Management and Research. 2016;**34**:316-326. DOI: 10.1177/0734242X16628979

[24] Guerrero AB, Aguado PL, Sánchez J, Curt MD. GIS-based assessment of banana residual biomass potential for ethanol production and power generation: A case study. Waste and Biomass Valorization. 2016;**7**:405-415. DOI: 10.1007/s12649-015-9455-3

[25] Comber A, Dickie J, Jarvis C, Phillips M, Tansey K. Locating bioenergy facilities using a modified GIS-based location-allocation-algorithm: Considering the spatial distribution of resource supply. Applied Energy. 2015;**154**:309-316. DOI: 10.1016/j.apenergy.2015.04.128

[26] Instituto Nacional de Estadística (INE). Chile: división político-administrativa y censal. Santiago, Chile; 2001. p. 222

[27] Ministerio del Medio Ambiente (MMA), Registro de emisiones y transferencia de contaminantes (RETC). 2012. Available on: http://www.mma.gob.cl/retc/1279/article-42161.html# [Accessed: 15-09-2018]

[28] Ministerio de Desarrollo Social, Base de datos de la encuesta de caracterización socioeconómica nacional (CASEN). 2012. Available on: http//observatorio.ministeriodesarrollosocial.gol.cl/basededatos.php [Accessed: 15-09-2018]

[29] Instituto Nacional de Estadística (INE). Catastro de viviendas electrificadas a escala distrital; Santiago, Chile; 2002

[30] Herrera-Seara MA, Dols FA, Zamorano M, Alameda-Hernández E. Optimal location of a biomass power plant in the province of Granada analyzed by multi-criteria

evaluation using appropriate Geographic Information System according to the Analytic Hierarchy Process, International Conference on Renewable Energies and Power Quality (ICREPQ'10); 23-25 March; Granada, Spain; 2010. pp. 813-818

[31] Andrade M, Pérez A, Alfonso D, Perpiña C, Tamayo Y, Rojas L, Armas R, Gámez A, Peñalvo E. Metodología para el aprovechamiento energético de biomasa en Cuba, 2009. Available on: http://bibdigital.epn.edu.ec/bitstream/15000/9382/2/P40.pdf [Accessed: 21-10-2018]

[32] Instituto Nacional de Estadística (INE). Chile: Ciudades, Pueblos, Aldeas y Caseríos. Santiago, Chile. 2005. p. 300

[33] Capilla JAJ, Carrión JA, Alameda-Hernández E. Optimal site selection for upper reservoirs in pump-back systems, using geographical information systems and multicriteria analysis. Renewable Energy. 2016;**86**:429-440. DOI: 10.1016/j.renene.2015.08.035

[34] Bili A, Vagiona DG. Use of multicriteria analysis and GIS for selecting sites for onshore wind farms: The case of Andros Island (Greece). European Journal of Environmental Sciences. 2018;**8**(1):5-13. DOI: 10.14712/23361964.2018.2

[35] Carrión JA, Estrella AE, Dols FA, Toro MZ, Rodríguez M, Ridao AR. Environmental decision-support systems for evaluating the carrying capacity of land areas: Optimal site selection for grid-connected photovoltaic power plants. Renewable and Sustainable Reviews. 2008;**12**(9):2358-2380. DOI: 10.1016/j.rser.2007.06.011

[36] Uyan M. GIS-based solar farms site selection using analytic hierarchy process (AHP) in Karapinar region, Konya/Turkey. Renewable and Sustainable Reviews. 2013;**28**:11-17. DOI: 10.1016/j.rser.2013.07.042

[37] Palma R, Jiménez G, Alarcón I. Las energías renovables no convencionales en el mercado eléctrico chileno. Fundación para la Transferencia Tecnológica (UNTEC) y Universidad de Chile. Santiago, Chile; 2009. 121 p

[38] Daniel J, Vishal NVR, Albert B, Selvarsan I. Evaluation of the significant renewable energy resources in India using analytical hierarchy process. In: Ehrgott M, Naujoks B, Stewart TJ, Wallenius J, editors. Multiple Criteria Decision Making for Sustainable Energy and Transportation Systems. Berlin: Springer-Verlag; 2010. pp. 13-26. DOI: 10.1007/978-3-642-04045-0 2

[39] Saaty T. Decision making with the analytic hierarchy process. International Journal of Services Sciences. 2008;**1**(1):83-98. DOI: 10.1504/IJSSci.2008.01759

[40] Salazar G, Irarrázaval F, Fonck M. Ciudades intermedias y gobiernos locales: desfases escalares en la Región de La Araucanía, Chile. EURE. 2017;**43**(130):161-184. DOI: 10.4067/s0250-71612017000300161

[41] Opazo J. The politics of system innovation for emerging technologies: Understanding the uptake of off-grid renewable electricity in rural Chile [thesis]. Falmer, Brighton, United Kingdom: University of Sussex; 2014

[42] Geneletti D. Assessing the impact of alternative land-use zoning policies on future ecosystem services. Environmental Impact Assessment Review. 2013;**40**:25-35. DOI: 10.1016/j.eiar.2012.12.003

[43] Romanelli JP, Silva LGM, Horta A, Picoli RA. Site selection for hydropower development: A GIS-based framework to improve planning in Brazil. Journal of Environmental Engineering. 2018;**144**(7):04018051. DOI: 10.1061/(ASCE)EE.1943-7870.0001381

[44] Mu E, Pereyra-Rojas M. Practical Decision Making. An Introduction to the Analytic Hierarchy Process (AHP). SpringerBriefs in Operations Research. Vol. 111. Switzerland: Springer International Publishing; 2017

[45] Alarcón DA. App for making decisions based on the analytical hierarchy process. Fides et Ratio. 2018;**15**(15):87-110

[46] Azócar L, Hermosilla N, Gay A, Rocha S, Díaz J, Jara P. Brown pellet production using wheat straw from southern cities in Chile. Fuel. 2019;**237**:823-832

Chapter 8

Sorghum an Important Annual Feedstock for Bioenergy

Bushra Sadia, Faisal Saeed Awan, Fozia Saleem, Ali Razzaq and Bushra Irshad

Abstract

Plant-based renewable biofuels guarantee sustainable solutions to food and energy demands. High-biomass C4 grasses including sugarcane, corn, and sorghum are potential candidates for bioenergy. Among these, sorghum enjoys the status of a highly diverse food, feed, and biofuel source worldwide. The natural attributes like abiotic stress tolerance, diverse genetic base, viable seed industry, and sound breeding system make sorghum a perfect candidate for establishing an efficient and low-cost biofuel industry. Scientists are exploring ways to exploit forage, sweet, and biomass sorghums as climate-smart energy crops. In this context, conventional breeding has played a significant role in developing high-yielding sorghum varieties. For biomass sorghum, stem compositional analysis helps screen low lignin and high polysaccharide types as feedstocks for biofuels. Recent tools of phenomics, genomics, proteomics, and genome editing are key players of designing eco-friendly bioenergy sorghum. Here, we report stem compositional analysis and proteomics-based evaluation of USDA sorghum germplasm as a baseline to develop sorghum as a biofuel feedstock.

Keywords: fossil fuels, feedstock, lignocellulosic biomass, C4 crops, genome, proteomics, hemicellulose, SSR

1. Introduction

In the present-day-global-warming era and with ever-increasing number of automobiles on the roads, fossil fuel reserves are going to be scarce and depleted over next few decades. In order to provide a safer environment to our future generations, we need to use energy wisely and economically and look for alternative fuel sources like biofuels, derived from crops and their waste products [1].

Biofuels are considered zero net emitters as they use atmospheric carbon dioxide for their growth and afterward release the same when burnt in the vehicles. The biofuels are generally classified as "conventional" (the first generation) and "advanced" biofuels (the second-, third-, and fourth-generation biofuels). Biodiesel and bioethanol are categorized as first-generation biofuels. These are produced from food crops rich in higher fermentable carbohydrate level. The second-generation biofuels are most commonly extracted from switchgrass, jatropha, miscanthus, and the residues of food crops. Often, industrial wastes are also used for the production of second-generation biofuel. The biofuels extracted from algae are classed as third generation. Major crops used for the production of biofuel are sugarcane, corn, wheat, sorghum, sugar beet, and cassava [2].

The choice of the most efficient biofuel depends upon its life cycle analysis, climatic, and economic factors. Moreover, its transportation cost to refinery, price of biofuel, and greenhouse gases also matter. Plant-based feedstocks for biofuels include crops like corn, sugarcane, soybean, poplar, sorghum, switch grass, etc. The cost-effective biofuel production depends upon the exploitation of high-yielding energy crops. Designing climate-smart energy crop with optimized composition to suit the growers, consumers, and industry needs is the backbone of cost-competitive biofuel industry. C4 grasses provide a perfect fit to this definition owing to higher photosynthetic rate, productivity, and broader genetic base of germplasm. Sorghum is a short duration crop of about 3–4 months and produces higher biomass yield with less inputs. These characteristics make sorghum a popular biofuel feedstock [3]. Sorghum has different end-use types including biomass, forage, sweet, and grain sorghums. Energy sorghum including biomass and sweet type varieties is the most efficient and climate-smart feedstock being able to grow with less inputs on marginal lands under harsh climatic conditions and having ability to utilize more sunlight [4–6].

It has diverse germplasm owing to extensive breeding and natural selection [7]. Sorghum is a crop of subsistence worldwide, the fifth most important cereal crop and an important component of poultry industry [8]. It is very responsive to biotechniques ranging from simple in vitro culture to transgenics, cisgenics, and genome-editing technologies. However, the outcrossing of sorghum with its weedy relatives has prevented regulation of GM technology in this crop. All above-ground parts of sorghum, starch, sugar, or stem biomass are utilized for the first- and second-generation biofuel production [9]. Though sweet sorghum has been widely used as a biofuel source, biomass sorghum has also been recently recognized as a promising feedstock for cellulosic ethanol production. This sorghum type usually has stem higher than 5 m, more number of leaves, fibrous roots, greater potential for vegetative growth, and is suitable for mechanization [10]. Besides producing second-generation ethanol, biomass sorghum also releases energy during biomass combustion [11]. It is a good substitute to corn and sugarcane with additional benefit of less water consumption. It is an annual grass having higher dry matter yield like perennial crops but in less duration, thus facilitating cheaper crop rotation. Recent wide scale applications of omics approaches like phenomics, genomics, proteomics, and metabolomics are enhancing the efficiency of sorghum breeding processes. Being an important element of system biology approach, omics analysis dissects the association between genes and proteins within diverse phenotypes. Genome analysis further refines this integration. Sorghum yields fuel and chemicals form sugars and cell wall biopolymers. Sorghum is a widely grown summer forage of Pakistan, while its biofuel potential is yet to be explored in the country. Information on sorghum stem quality traits is vital for designing eco-friendly biofuel source. Present study intended to demonstrate the basis of morphological characterization of 24 USDA sorghum genotypes selected under Pakistan conditions. These genotypes were subjected to proximate analysis to measure stem quality traits like crude protein, ash contents, neutral detergent fiber, acid detergent fiber, hemicellulose, cellulose, and acid detergent lignin. Translational analysis indicated a unique band of 56.1 kDa in 12 out of 24 genotypes. This uncharacterized protein is supposed to be translated by Dw1 gene (Sobic.009G229800) comprising of 510 amino acids and controls the internodal length in sorghum. In this chapter, stem composition evaluation and proteomics-based recent research involving USDA sorghum germplasm is reported in order to screen promising energy-type sorghum.

2. Analyzing sorghum biomass potential

2.1 Phenotyping biomass sorghums

Sorghum biomass is influenced by genetic and environmental factors [12]. The identification of variation in phenotypic, genetic, structural, and physiological characters of energy sorghum is vital to its improvement. Sorghum biomass improvement model relies on integrating several genomic-assisted techniques with phenomics approaches. Common field-based selection of high biomass sorghum depends upon characterizing biomass-related morphological traits like days to flowering (days after sowing), plant height, fresh biomass yield, dry matter, and dry matter yield, plant height, stalk diameter, leaf number, leaf width, leaf length, leaf angle and leaf area index, etc. [13]. Several studies report on morphological diversity assessment of sorghum for biomass traits in the field environment [14, 15].

Accurate and comprehensive phenotypic data are the baseline to elucidate genetic mechanisms underlying complex quantitative biomass traits. Since biomass-related traits are measured via destructive sampling, recording morphological data during the entire growing period of energy sorghum is possible only once. Manual, nondestructive sampling for these traits over complete development of sorghum is impossible. As compared to relatively cheaper technologies of genomic selection, association mapping and GWAS, reliable phenotyping is laborious and expensive. About 20 years back when genotyping techniques were fast advancing, improving phenotyping approaches was completely ignored. Recently, there has been a growing interest in developing effective sorghum phenotyping methods. The work started with optimizing high-throughput phenotyping systems for model plants under controlled environments. Later on, field-based phenotyping platforms were devised for short stature crops [16]. In the last 5 years, different approaches have been excogitated with promising capabilities of recording sorghum phenology in field environments. Some of these include various UAS platforms [17, 18], field-based robotic phenotyping system [19], unmanned aerial system [20], ultrasonic sensors [21], the light detection and ranging (LiDAR) [22], the time of flight cameras [23], tomography imaging [24], Kinect v2 camera [21], RGB and NIR imaging [25], and Phenobot 1.0 [26]. The next-generation phenomics tools generate enormous amount of data that are being translated via machine- learning statistical approaches into trait descriptions, relevant to sorghum breeders [27].

2.2 Analyzing biomass stem composition

The composition of biomass derived from forage, grain, and sweet sorghums has been well characterized [28]. The research on exploiting forage sorghum as biofuel was initiated in 1980s, which led to the development of photoperiod-sensitive-energy sorghum hybrids [29]. These are high biomass yielders [30]. Being relatively a recent introduction, the stem composition knowledge of energy sorghum is still limited. Up till now, a majority of research on sorghum biomass feedstock has focused more on improving yield than the quality components. So, there is a need to accurately conduct the biochemical analysis, since stem composition is the basic element influencing biofuel yield.

Plant cell walls are the main constituents of biomass that provide strength and limited plasticity to cell. The cell wall serves as a tough physical barrier, protecting interior of the cell against biotic and abiotic stresses. It is a multilayered structure composed of polysaccharides and proteins, which are important contributors of biofuel quality and energy conversion processes. The polysaccharides are cellulose (a polymer of glucose), pectic compounds (polymers of galacturonic acid), and hemicellulose (a polymer of a variety of sugars including xylose, arabinose, and mannose). Cellulose is the largest source of glucose for biofuels. Glucuronoarabinoxylan (GAX) hemicellulose complex is linked to lignin. Since lignin component of plant cell wall provides structure, it cannot be converted to carbohydrates and hence is recalcitrant to conversion protocols. Likewise, ash content also reduces biomass to biofuel conversion reaction. Certain constituents of cell wall are water soluble like sugars, proteins, amino acids, mixed-linkage glucans, and phenolic glycosides, whereas chlorophyll, lipids, and waxes are water-insoluble ingredients that need ethanol extraction.

Different studies have reported various approaches for compositional analysis of energy sorghum leaves and stem. In some sorghum genotypes, proportion of cellulose can vary between 27 and 52%, while the range of hemicellulose content is 17–23% and lignin content is 6.2–8.1% [31, 32]. Along with the biomass yield, low lignin, high cellulose, and hemicellulose contents are also the desirable selection attributes for energy sorghum genotypes [33]. Such sorghums exhibit wide variations in biomass composition [34]. Now a days, near-infrared spectroscopic (NIR) analysis is routinely used for high-throughput computation of biomass composition [28].

Cellulosic bioethanol production requires three main steps: pretreatment, hydrolysis, and fermentation [35] (**Figure 1**). Pretreatment is performed to fractionate lignocellulose into different components via physical (boiling, steaming, and ultrasonication), chemical (acid, alkali, salts, etc.), physiochemical (ammonium fiber explosion or AFEX), and biological methods (bacteria and fungi). It increases porosity and surface area of the substrate. During hydrolysis, nonstructural carbohydrates are degraded in to sugars. Enzyme-based hydrolysis is preferred over acid hydrolysis being a mild and cost-effective process.

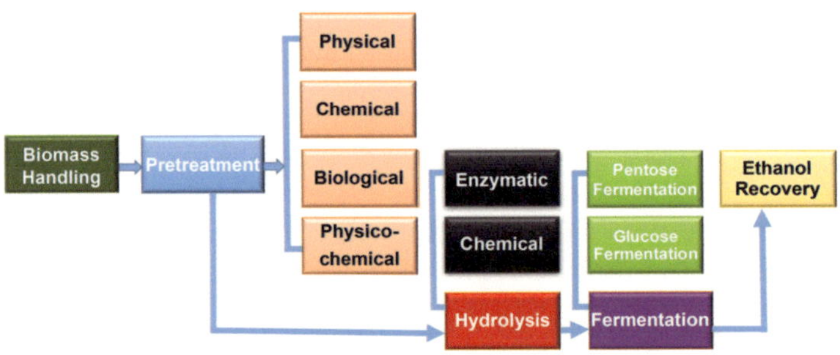

Figure 1.
Flow chart of sorghum cellulosic ethanol production process.

The process of fermentation proceeds under liquid or solid state in the presence of bacteria or yeast [36]. In a recent study, 24 sorghum genotypes (**Table 1**) were subjected to stem compositional analysis [37]. These genotypes had previously been selected on the basis of morphological traits [38].

The dried stem samples of these genotypes were grinded and used for measuring crude protein (%), ash contents (%), neutral detergent fiber (NDF %), acid detergent fiber (ADF %), hemicellulose (%), cellulose (%), and acid detergent lignin (ADL %), using the respective formulae:

$$\text{Crude protein } \% = \frac{0.1\,\text{N}\,H_2SO_4 \times 100 \times 6.25 \times (0.0014 \times \text{total diluted volume})}{\text{Weight of sample} \times \text{diluted digested material (ml)}} \tag{1}$$

Sr. #	Genotype #	Sr. #	Genotype #
1.	PI-609239-01-SD	13.	PI-329875-03-SD
2.	PI-620625-01-SD	14.	PI-330039-02-SD
3.	PI-648173-01-SD	15.	PI-330022-01-SD
4.	PI-648187-01-SD	16.	PI-456415-03-SD
5.	PI-454464-03-SD	17.	PI-329488-02-SD
6.	PI-570039-02-SD	18.	PI-155871-02-SD
7.	PI-525981-01-SD	19.	PI-457393-02-SD
8.	PI-329569-01-SD	20.	PI-329480-02-SD
9.	PI-583832-02-SD	21.	PI-303658-02-SD
10.	PI-329733-01-SD	22.	PI-303656-01-SD
11.	PI-456441-03-SD	23.	NSL-54978
12.	PI-329471-02-SD	24.	PI-257595-01-SD

Table 1.
Sorghum genotypes used for stem compositional analysis.

$$\text{Ash\%} = \frac{\text{Weight of ash} \times 100}{\text{Weight of sample}} \tag{2}$$

$$\text{NDF\%} = \frac{(\text{Weight of crucible residue}) - \text{Weight of crucible} \times 100}{\text{Weight of sample}} \tag{3}$$

$$\text{ADF\%} = \frac{(\text{Weight of crucible} + \text{ADF residue}) - \text{Weight of crucible} \times 100}{\text{Weight of sample}} \tag{4}$$

$$\text{Hemicellulose \%} = \frac{(\text{NDF} - \text{ADF}) - \text{Weight of crucible} \times 100}{\text{Weight of sample}} \tag{5}$$

$$\text{Cellulose\%} = \frac{(\text{Weight of crucible} + \text{ADF residue}) - \text{Weight.of crucible} + \text{residue after 24 } NH_2SO_4 \times 100}{\text{Weight of sample}} \tag{6}$$

$$\text{Lignin/ADL (\%)} = \frac{(\text{Weight of crucible} + \text{residue of cellulose}) - \text{Weight of crucible} + \text{residue after combustion} \times 100}{\text{Weight of sample}} \tag{7}$$

Statistical analysis indicated highly significant variations among all sorghum genotypes for crude protein, ash contents, NDF, ADF, ADL, hemicellulose, and cellulose contents (**Table 2**).

PCA analysis of different biochemical traits indicated three principle components (PC1, PC2, and PC3) having Eigen values greater than 1 (**Table 3**). The cumulative variability of three PCs was 82.94% for the studied genotypes. The total variability in traits shared by three PCs was 37.48, 27.37, and 18.096%, respectively. Different biomass-related traits added more than 34% of variation factor in PC1 such as: ash contents (43.7%), ADL (47.6%), cellulose (45.5%), hemicelluloses (37.4%), and NDF (48.5%). PC1 showed weak and positive correlation with crude protein (0.000%) and ADF (0.012%). The PC2 contributed for 27.37% of total variability. PC2 showed positive and strong correlation with the traits such as ADL (38.4%), ADF (50.1%), and cellulose (46.8%). Weak and negative correlation was

Eigen vectors	PC1	PC2	PC3
AC	0.437	−0.218	0.373
ADL	0.476	0.384	0.134
ADF	0.012	0.501	−0.191
C	0.455	0.468	0.097
CP	0.000	−0.188	0.777
HC	0.374	−0.481	−0.293
NDF	0.485	−0.261	−0.328
Eigen value	2.623	1.916	1.267
Variability %	37.476	27.371	18.096
Cumulative %	37.476	64.847	82.943

PC, principle component; SD, standard deviation; CV, coefficient of variation; AC, ash contents; ADL, acid detergent lignin; ADF, acid detergent fiber; C, cellulose; CP, crude protein; HC, hemicellulose; NDF, neutral detergent fiber.

Table 2.
Principle component analysis (PCA) related to biomass traits in sorghum.

Variables	Minimum	Maximum	Mean	SD	CV (%)
CP	4.927	10.927	7.808	1.414	1.37
AC	5.217	19.470	12.418	3.877	2.39
NDF	54.633	81.500	63.947	6.411	2.29
ADF	26.167	54.500	34.410	6.994	4.34
ADL	1.500	8.000	3.160	1.316	14.17
HC	22.087	44.150	31.419	5.981	1.64
C	29.000	57.167	39.250	7.331	2.03

SD, standard deviation; CV, coefficient of variation; AC, ash contents; ADL, acid detergent lignin; ADF, acid detergent fiber; C, cellulose; CP, crude protein; HC, hemicellulose; NDF, neutral detergent fiber.

Table 3.
Descriptive statistics for quantitative traits of sorghum germplasm.

observed for ash contents (21.8%), crude protein (18.8%), hemicellulose (48.1%), and NDF (26.1%). Crude protein and ash contents showed 77.7 and 37.3% of the factor variations in PC3, respectively.

Biplot analysis described that variables were greatly obliged as vectors; comparative length of the vector was distinguished as the relative proportion of the variability in each variable. The traits like ADL and CP, which were plotted near the central point, showed more similarities, while cellulose, ADF, NDF, and HC displayed more variability (**Figure 2**). Significant characters such as ADL, ADF, and cellulose were located at positive and positive coordinate region in biplot. Traits like AC, NDF, and HC were allocated at negative coordinate (**Figure 2**). Variability in the traits explains the variations among genotypes, which can be used in sorghum breeding plan effectively. Correlation analysis among biofuel-related stem compositional traits indicated that concentration of protein and lignin contents showed negative interaction with cellulose and hemicelluloses (**Table 4**). It showed that significant genetic variability is present among 24 sorghum genotypes. In sorghum, cellulose and hemicellulose contents play significant role in biofuel quality. For fiber

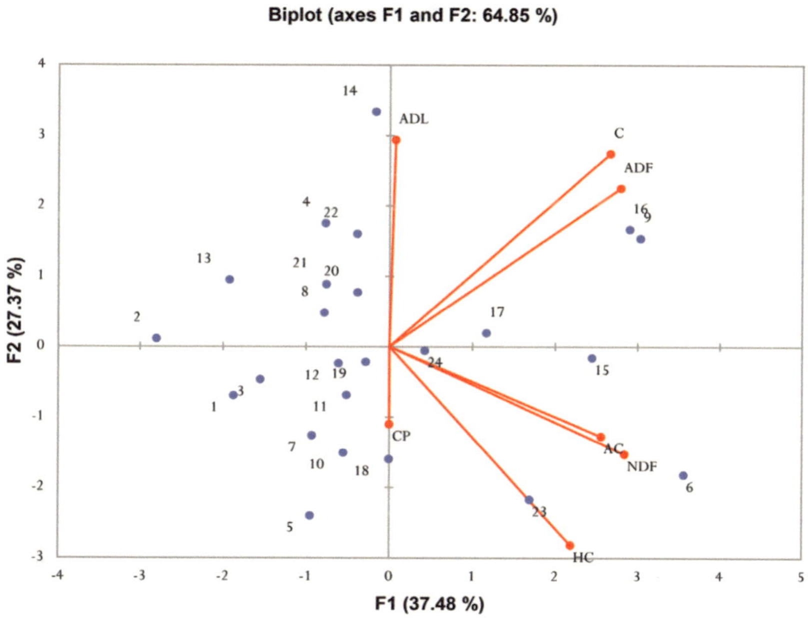

Figure 2.
PCA grouping of 24 USDA sorghum genotypes using quantitative traits.

Traits	CP	AC	NDF	ADF	ADL	HC	C
CP	1	0.347*	−0.128	−0.066	−0.182	−0.051	−0.080
AC	0.347	1	0.473*	0.362*	−0.173	0.431*	0.311*
NDF	−0.128	0.473	1	0.293*	−0.033	0.802**	0.289*
ADF	−0.066	0.362	0.293	1	−0.153	0.067	0.955**
ADL	−0.182	−0.173	−0.033	0.153	1	−0.313	0.335*
HC	−0.051	0.431	0.802	0.067	−0.313	1	−0.016
C	−0.080	0.311	0.289	0.955	0.335	−0.016	1

*Normal correlation.
**Strong correlation.

Table 4.
Correlation coefficients of various traits of sorghum genotypes.

analysis, NDF, ADL, and ADF are generally used as standard quality testing techniques [39], while lignin concentration markedly affects the efficiency of hydrolysis [40].

Study reports that by increasing the level of lignin, cellulose and hemicellulose concentrations decreased. The genetic relationships among 24 genotypes were identified through construction of dendrogram on the basis of similarity matrix utilizing the UPGMA algorithm (**Figure 3**). The genotypes were grouped into two main clusters: only two genotypes (PI-583832-02-SD and PI-456415-03-SD) were present in subcluster-1, while the subcluster-2 was divided into smaller groups. The genotypes PI-570039-02-SD, PI-330022-01-SD, and NSL-54978 were grouped together and showed some distinctness from rest of the members of the group, whereas the maximum genetic relatedness was found among genotypes PI-329569-01-SD and PI-303658-02-SD followed by genotypes PI-329733-01-SD, PI-525981-01-SD, PI-303656-01-SD, and PI-648187-01-SD. The genotypes

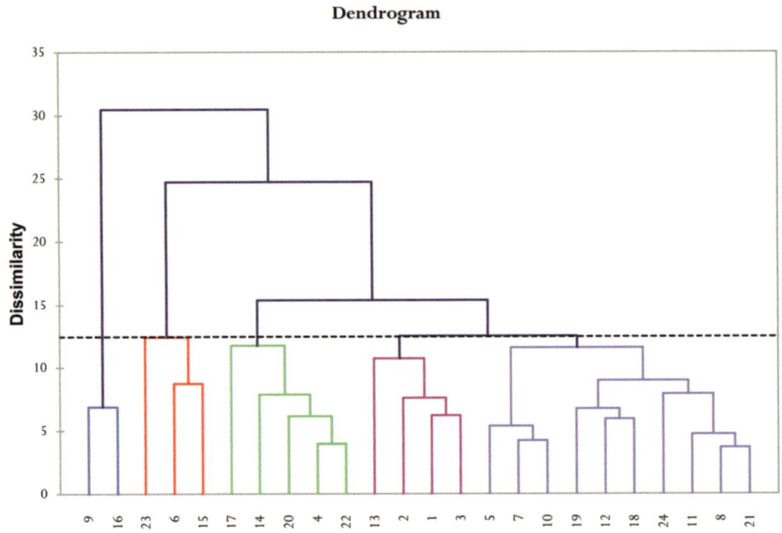

Figure 3.
Classification of 24 sorghum genotypes using UPGMA cluster analysis.

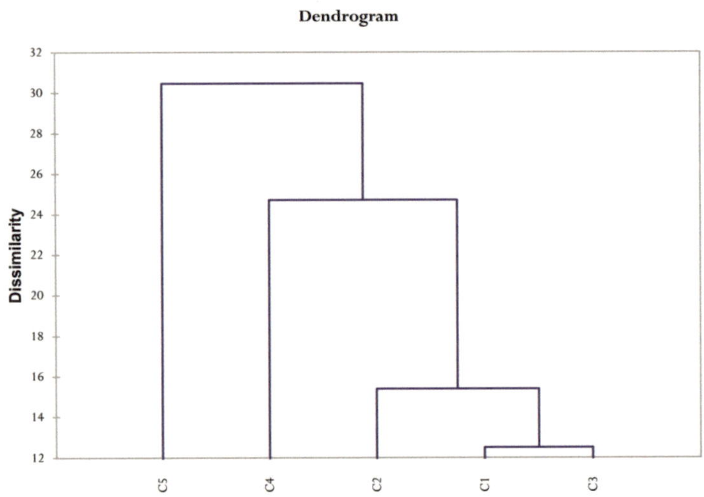

Figure 4.
Cladogenesis studies using homology-based classification of 24 sorghum genotypes.

PI-583832-02-SD and PI-329733-01-SD were also found genetically distinct from rest of the genotypes used in the study (**Figure 4**). Variance decomposition for optimal classification showed that there were 23.41 and 76.59% variances present within and between classes, respectively.

The sorghum germplasm with less lignin and protein contents is desirable for biofuel production. Sorghum genotype PI-609239-01-SD had maximum value of NDF (83.5%) and ash contents (19.5%), while genotype PI-303658-02-SD exhibited the maximum value (57.5%) of cellulose content.

Though sorghum is viewed as a cheap source of biofuel being able to grow on marginal lands, few studies have indicated the lower biofuel potential of energy sorghums grown on marginal lands than the crop land [41]. Hence, screening of energy sorghum having stress tolerance, with efficient production technology and conservation tillage practices, is the key element of sustainable commercial production of energy sorghum [5].

2.3 Transcriptional and translational analyses of sorghum biomass

The mysterious relationship between phenotype and genotype can be revealed by applying various biotechnological approaches such as proteomics, transcriptomics, and metabolomics [42]. In transcriptomics, a huge set of gene libraries can be established by employing different techniques of bioinformatics and next-generation sequencing [43]. Over the last decade, expression profiling experiments for genome-wide investigation in sorghum have been carried out to analyze responses to numerous abiotic and biotic stresses, to determine tissue-specific and genotype-specific gene expression motifs, and to disclose the genetic modification and expression divergence between different sorghum varieties.

RNA-seq technology for expression profiling has been applied in sorghum to study different gene functions [44]. This technique gives a precise assessment of gene expression at different stages of sorghum plant development [45].

Proteomics offers the set of the most efficient tools for recognition, assessment, and quantification of unique proteins. Our recent study [44] merged transcriptomic and proteomic approaches for screening sorghum germplasm best suited for bioenergy and for comparative analysis of protein expression of elite sorghum germplasm. The study was based on 24 USDA sorghum genotypes selected for biomass potential in the field experiments, which is already reported in this chapter [37]. For translational analysis, 12 out of 24 selected genotypes were divided into three groups based on stem height, since height is directly correlated with biomass in sorghum. Four short stature genotypes were chosen as negative control (**Table 5**).

Sr. #	Genotypes	Height-based groups
1.	NSL-54978	Tall
2.	PI-456441-03-SD	
3.	PI-525981-01-SD	
4.	PI-303656-01-SD	
5.	PI-457393-02-SD	Medium
6.	PI-583832-02-SD	
7.	PI-620625-01-SD	
8.	PI-456415-03-SD	
9.	PI-648187-01-SD	Small
10.	PI-609239-01-SD	
11.	PI-330039-02-SD	
12.	PI-329733-01-SD	
13.	PI-643630-01-SD	Negative control
14.	PI-643735-03-SD	
15.	PI-643581-01-SD	
16.	PI-642993-01-SD	

Table 5.
Sorghum genotypes and their respective groups based on height.

The *in vitro*-germinated, 15-day-old sorghum seedlings were used for protein extraction. Sodium dodecyl sulfate polyacrylamide gel electrophoresis (SDS-PAGE) revealed diverse banding pattern of proteins ranging in size from 14.9 to 124 kDa with different expression levels in all studied genotypes (**Table 6**).

Genotype	Protein weight (kDa)								
NSL-54978	124	97.6		64	56.1	40.5	38.7	32	14.9
PI-456441-03-SD	124	97.6		64	56.1	40.5	38.8	32	14.9
PI-525981-01-SD	124	97.6	71	64	56.1	40.5	38.8	32	14.9
PI-303656-01-SD	124	97.6		64	56.1	40.5	38.8	32	14.9
PI-457393-02-SD			71	64	56.1	40.5	38.8	32	
PI-583832-02-SD		97.6	71	64	56.1	40.5	38.8	32	
PI-620625-01-SD			71	64	56.1	40.5	38.8	32	
PI-456415-03-SD			71	64	56.1	40.5	38.8	32	
PI-648187-01-SD		97.6	71	64	56.1	40.5	38.8	32	
PI-609239-01-SD			71	64	56.1	40.5	38.8	32	
PI-330039-02-SD		97.6		64	56.1	40.5	38.8	32	
PI-329733-01-SD		97.6		64	56.1	40.5	38.8	32	
PI-643630-01-SD		97.6		64		40.5		32	
PI-643735-03-SD		97.6		64		40.5		32	
PI-643581-01-SD				64		40.5		32	
PI-642993-01-SD				64		40.5		32	

Table 6.
SDS-PAGE-based banding pattern of various proteins in sorghum genotypes.

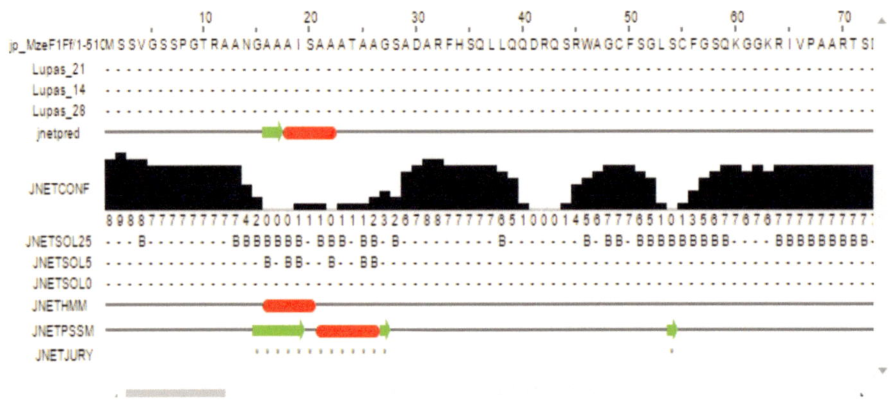

Figure 5.
Secondary structure prediction of SORB1_3009G229800 protein responsible for stem internodal length.

SDS-PAGE showed nine different bands in 12 selected sorghum genotypes. The banding pattern of four negative controls was different from the selected ones, which revealed low expression of proteins. The study showed a unique band of 56.1 kDa present only in all selected genotypes. This band represents a hypothetical protein Sobic.009G229800, which has 510 amino acids (**Figures 5** and **6**) and controls the internodal length of stem in sorghum, which is why short-stature sorghum genotypes were devoid of this protein.

Height is positively correlated with biomass production [46] and is reported to be independent of stem structural composition like cellulose, hemicellulose, and lignin contents [47]. The Quantitative trait loci (QTL) for total dry biomass has been found to be localized with height QTLs [48]. Hence, breeders aim for taller

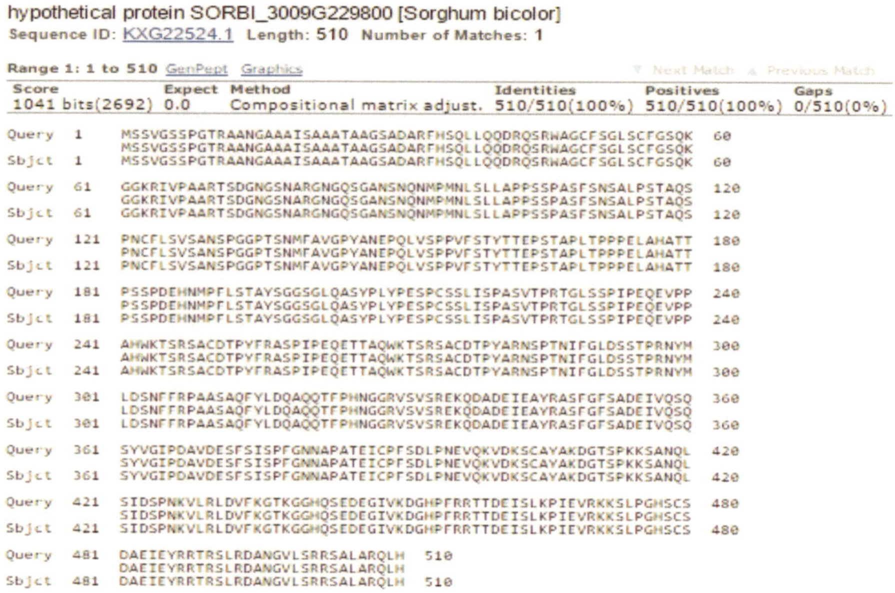

Figure 6.
Blast result for confirming the SORB1_3009G229800 protein against NCBI database.

Names and taxonomy	
Protein	Uncharacterized protein
Gene	SORBI_009G229800
Organism	*Sorghum bicolor*
Taxonomic identifier	4558 [NCBI]
Proteomes	UP000000768
Chromosome	9
Sequence databases	CM000768 Genomic DNA Translation KXG22524.1
Last sequence update	November 2, 2016

Table 7.
Profile of SORBI_3009G229800 protein translated from Dw1 *gene and upregulated in top sorghum genotypes.*

genotypes in sorghum biomass improvement plans. Chromosomes six, seven, and nine carry QTLs for height in sorghum. This protein (Sobic.0 09G229800) is considered to be translated from Dw1, a gene greatly conserved in plants (**Table 7**). Earlier reports showed that *Dw1* enhances the internodal length and weight of sorghum plant [49] and is in turn important for plant biomass production.

3. Conclusion

Energy sorghum is considered to be a promising biofuel feedstock to counteract the depleted fossil fuel reserves. To keep pace with fast progressing sorghum genomics, recent phenomics tools have been evolved that are more efficient than traditional laborious field-based manual phenotyping methods. This chapter describes the results of recent studies involving 24 selected biomass sorghums. The genotypes with low lignin, high cellulose, and hemicellulose components have been identified. Furthermore, with the help of translational analysis, an uncharacterized protein (Sobic.009G229800) is identified in tall sorghum genotypes. It regulates plant height by altering the length of internodes. Sorghum feedstock's stem compositional analysis, genomics, phenomics, and proteomics are enabling technologies extensively used by sorghum researchers for selection of elite sorghum germplasm with biofuel potential.

Author details

Bushra Sadia[1,2*], Faisal Saeed Awan[1], Fozia Saleem[1], Ali Razzaq[1] and Bushra Irshad[1]

1 Centre of Agricultural Biochemistry and Biotechnology (CABB), University of Agriculture, Faisalabad (UAF), Pakistan

2 US-Pakistan Centre for Advanced Studies in Agriculture and Food Security (USPCAS-AFS), University of Agriculture, Faisalabad (UAF), Pakistan

*Address all correspondence to: bushra.sadia@uaf.edu.pk

References

[1] Sanscartier D, Deen B, Dias G, Maclean H, Dadfar H, Mcdonald I, et al. Implications of land class and environmental factors on life cycle GHG emissions of Miscanthus as a bioenergy feedstock. GCB Bioenergy. 2014;**6**: 401-413

[2] Janda K, Kristoufek L, Zilberman D. Biofuels: Policies and impacts. Agricultural Economics. 2012;**58**: 372-386

[3] Ratnavathi C, Suresh K, Kumar BV, Pallavi M, Komala V, Seetharama N. Study on genotypic variation for ethanol production from sweet sorghum juice. Biomass and Bioenergy. 2010;**34**: 947-952

[4] Xie GH. Progress and direction of non-food biomass feedstock supply research and development in China. Journal of China Agricultural University. 2012;**17**:1-19

[5] Olson SN, Ritter K, Rooney W, Kemanian A, McCarl BA, Zhang Y, et al. High biomass yield energy sorghum: Developing a genetic model for C4 grass bioenergy crops. Biofuels, Bioproducts and Biorefining. 2012;**6**: 640-655

[6] Sher A, Hassan FU, Ali H, Hassan W. Seed rate and nitrogen application effects on production and brix value of forage sorghum cultivars. Grassland Science. 2016;**62**:119-127

[7] Paterson A, Bowers JE, Bruggmann R, Dubchak I, Grimwood J, et al. The Sorghum bicolor genome and the diversification of grasses. Nature. 2009; **457**:551-556

[8] Boyles RE, Brenton ZW, Kresovich S. Genetic and genomic resources of sorghum to connect genotype with phenotype in contrasting environments. The Plant Journal. 2019;**97**(1):19-39. DOI: 10.1111/tpj.14113

[9] Cotton J, Burow G, Acosta-Martinez V, Moore-Kucera J. Biomass and cellulosic ethanol production of forage sorghum under limited water conditions. Bioenergy Research. 2013;**6**: 711-718

[10] Venuto B, Kindiger B. Forage and biomass feedstock production from hybrid forage sorghum and sorghum–sudangrass hybrids. Grassland Science. 2008;**54**:189-196

[11] da Silva MJ, Carneiro PCS, Carneiro JES, Damasceno CMB, Parrella NNLD, Pastina MM, et al. Evaluation of the potential of lines and hybrids of biomass sorghum. Industrial Crops and Products. 2018;**125**:379-385

[12] Betts NS, Fox GP, Kelly AM, Cruickshank AW, Lahnstein J, Henderson M, et al. Non-cellulosic cell wall polysaccharides are subject to genotype × environment effects in sorghum (*Sorghum bicolor*) grain. Journal of Cereal Science. 2015;**63**:64-71

[13] Arshad SF, Sadia B, Awan FS, Jaskani MJ. Estimation of genetic divergence among sorghum germplasm of Pakistan through multivariate tools. IJAB. 2017;**19**:1099-1106

[14] Dossou-Aminon I, Loko YL, Adjatin A, Dansi A, Elangovan M, Chaudhary P, et al. Diversity, genetic erosion and farmer's preference of sorghum varieties [*Sorghum bicolor* (L.) Moench] growing in North-Eastern Benin. International Journal of Current Microbiology and Applied Sciences. 2014;**3**:531-552

[15] Safdar H. Application of microsatellites in genetic diversity analysis of USDA sorghum germplasm [MPhil dissertation]. Pakistan:

University of Agriculture, Faisalabad; 2018

[16] Araus JL, Cairns JE. Field high-throughput phenotyping: The new crop breeding frontier. Trends in Plant Science. 2014;**19**:52-61

[17] Anthony D, Elbaum S, Lorenz A, Detweiler C. On crop height estimation with UAVs. In: Burgard W, editor. IEEE/RSJ International Conference on Intelligent Robots and Systems; 14–18 Sept. 2014; Chicago. New York: IEEE; 2014. pp. 4805-4812

[18] Watanabe K, Guo W, Arai K, Takanashi H, Kajiya-Kanegae H, Kobayashi M. High-throughput phenotyping of sorghum plant height using an unmanned aerial vehicle and its application to genomic prediction modeling. Frontiers in Plant Science. 2017;**8**:421. DOI: 10.3389/fpls.2017.00421

[19] Bao Y, Tang L. Field-based robotic phenotyping for sorghum biomass yield component traits characterization using stereo vision. IFAC-PapersOnLine. 2016;**49**:265-270

[20] Pugh NA, Horne DW, Murray SC, Carvalho G, Malambo L, Jung J, et al. Temporal estimates of crop growth in sorghum and maize breeding enabled by unmanned aerial systems. Frontiers in Plant Science. 2017;**6**:820

[21] Wang X, Singh D, Marla S, Morris G, Poland J. Field-based high-throughput phenotyping of plant height in sorghum using different sensing technologies. Plant Methods. 2018;**14**:53

[22] Malamboa L, Popescua SC, Horneb DW, Pughb NA, Rooney WL. Automated detection and measurement of individual sorghum panicles using density-based clustering of terrestrial lidar data. ISPRS Journal of Photogrammetry and Remote Sensing. 2019;**149**:1-13

[23] McCormick RF, Truong SK, Mullet JE. 3D sorghum reconstructions from depth images identify QTL regulating shoot architecture. Plant Physiology. 2016;**172**:823-834

[24] Karve AA, Alexoff D, Kim D, et al. In vivo quantitative imaging of photoassimilate transport dynamics and allocation in large plants using a commercial positron emission tomography (PET) scanner. BMC Plant Biology. 2015;**15**:273

[25] Neilson H, Moloney M, Inzé D. Translational research: From pot to plot. Plant Biotechnology Journal. 2014;**12**: 277-285

[26] Salas Fernandez MG, Bao Y, Tang L, Schnable PS. A high-throughput, field-based phenotyping technology for tall biomass crops. Plant Physiology. 2017; **174**:2008-2022

[27] Perez-Sanz F, Navarro PJ, Egea-Cortines M. Plant phenomics: An overview of image acquisition technologies and image data analysis algorithms. GigaScience. 2017;**6**:1 gix092. DOI: 10.1093/gigascience/gix092

[28] McKinley BA, Olson SN, Ritter KB, et al. Variation in energy sorghum hybrid TX08001 biomass composition and lignin chemistry during development under irrigated and non-irrigated field conditions. PLoS One. 2018;**13**(4):e0195863. DOI: 10.1371/journal.pone.0195863

[29] Gill JR, Burks PS, Staggenborg SA, Odvody GN, Heiniger RW, Macoon B, et al. Yield results and stability analysis from the sorghum regional biomass feedstock trial. BioEnergy Research. 2014;**7**:1026-1034. DOI: 10.1007/s12155-014-9445-5

[30] Rooney WL, Blumenthal J, Bean B, Mullet JE. Designing sorghum as a dedicated bioenergy feedstock. Biofuels,

Bioproducts and Biorefining. 2007;**1**: 147-157

[31] Reddy BVS, Ramesh S, T Longvah T. Prospects of breeding for micronutrients and carotene-dense sorghums. International Sorghum and Millets Newsletter. 2005; **46**:10-14

[32] Bennett AS, Anex RP. Production, transportation and milling costs of sweet sorghum as a feedstock for centralized bioethanol production in the upper Midwest. Bioresource Technology. 2009;**100**:1595-1607

[33] Mahmood A, Honermeier B. Chemical composition and methane yield of sorghum cultivars with contrasting row spacing. Field Crops Research. 2012;**128**:27-33

[34] Damasceno R, Schaffert E, Dweikat I. Mining genetic diversity of sorghum as a bioenergy feedstock. In: McCann MC, Marcos SB, Carpita NC, editors. Plants and Bioenergy. Advances in Plant Biology. New York: Springer-Verlag; Vol. 4. 2014. pp. 81-102

[35] Fracasso A, Perego A, Amaducci S. Characterisation of ten commercial sorghum genotypes grown under water-limited conditions for bioenergy production in Mediterranean environment. Italian Journal of Agronomy. 2017;**12**:302-314. DOI: 10.4081/ija.2017.898

[36] Bensah EC, Mensah M. Chemical pretreatment methods for the production of cellulosic ethanol: Technologies and innovations. International Journal of Chemical Engineering. 2013;**1**:21

[37] Irshad B. Chemical and genetic diversity analyses of high biomass USDA sorghum (*Sorghum bicolor* M.) collections. [MPhil dissertation]. , Faisalabad: University of Agriculture; 2017

[38] Sadia B, Awan FS, Saleem F, Sadaqat HA, Arshad SF, Shaukat H. Genetic improvement of sorghum for biomass traits using genomics approaches. In: Rao MN, Soneji JR, editors. Advances in Biofuels and Bioenergy. IntechOpen. DOI: 10.5772/intechopen.73010

[39] Miron J, Solmon R, Adin G, Nir U, Nikbachat M, Yosef E, et al. Effects of harvest stage and re-growth on yield, ensilage and in vitro digestibility of new forage sorghum varieties. Journal of the Science of Food and Agriculture. 2006; **86**:140-147

[40] Fu C, Mielenz JR, Xiao X, Ge Y, Hamilton CY, et al. Genetic manipulation of lignin reduces recalcitrance and improves ethanol production from switchgrass. PNAS. 2011;**108**:3803-3808

[41] Li CX, Feng HS. A study on the adaptability of sweet sorghum planted in different altitudinal areas of the Qinghai plateau. Acta Prataculturae Sinica. 2013;**22**:51-59

[42] Turner MF, Heuberger AL, Kirkwood JS, Collins CC, Wolfrum EJ, Broeckling CD, et al. Non-targeted metabolomics in diverse sorghum breeding lines indicates primary and secondary metabolite profiles are associated with plant biomass accumulation and photosynthesis. Frontiers in Plant Science. 2016;7:953

[43] Van Bel M, Proost S, Van Neste C, Deforce D, Van de Peer Y, Vandepoele K. TRAPID: An efficient online tool for the functional and comparative analysis of de novo RNA-Seq transcriptomes. Genome Biology. 2013;**14**:R134

[44] Razzaq A. Transcriptional and translational analysis of bioenergy smart sorghum [MPhil dissertation].

Faisalabad: University of Agriculture; 2017

[45] Shimada S, Makita Y, Kuriyama-Kondou T, Kawashima M, Mochizuki Y, Hirakawa H, et al. Functional and expression analyses of transcripts based on full-length cDNAs of Sorghum bicolor. DNA Research. 2015;**22**:485-493

[46] Arshad SF. Identification of biomass-related QTLs in sorghum (*Sorghum bicolor* L. Moench) using association mapping [PhD dissertation]. Faisalabad: University of Agriculture; 2017

[47] Murray SC, Rooney WL, Mitchell SE, Sharma A, Klein PE, Mullet JE, et al. Genetic improvement of sorghum as a biofuel feedstock: II. QTL for stem and leaf structural carbohydrates. Crop Science. 2008;**48**:2180-2193

[48] Brown PJ, Rooney WL, Franks C, Kresovich S. Efficient mapping of plant height quantitative trait loci in a sorghum association population with introgressed dwarfing genes. Genetics. 2008;**180**:629-637

[49] Hilley J, Truong S, Olson S, Morishige D, Mullet J. Identification of Dw1, a regulator of sorghum stem internode length. PLoS One. 2016;**11**: e0151271

Chapter 9

Energy and Exergy Analysis of an Advanced Cookstove-Based Annular Thermoelectric Cogeneration System

Sakthivadivel Duraisamy, Manikandan Sundararaj, Kumaraselvan Raja, Ganesh Kumar Poongavanam and Iniyan Selvarasan

Abstract

This chapter deals with the energy and exergy analysis of the cookstove-based gasifier annular thermoelectric generator (GATEG). The vented waste heat is made available at the outer surface of the combustion chamber of an advanced micro-gasifier cookstove for added energy feed to the GATEG. This combined device has a competence to satisfy both cooking needs and micro-electrification of rural villages by a simultaneous recovery of heat energy and power (CHP) as cogeneration system. The power output (W), electrical energy efficiency (%) and exergy efficiency (%) of the proposed advanced micro-gasifier cookstove-based ATEG are 10 W, 6.78 and 15%, respectively. The maximum hot side wasted temperature without annulus gain is 275°C, which translates equivalent loss values as 7.64 W, 5.45 and 10.49%; this loss is higher than achievable minimum hot side temperature of 150°C on which this analytical chapter is drafted. This detailed study will be extremely useful to the designers of commercial biomass advanced micro-gasifier cookstove integrated ATEG systems.

Keywords: cookstove, energy, exergy, power, annular thermoelectric generator

1. Introduction

Clean sources of enhanced energy recovery offer the best mitigating solution to the economic, environmental and climate effects from the continued high consumption/utilisation of fossil fuels. Biomass resource is an excessive source of carbon-neutral renewable energy (RE) available far and worldwide and is a good source of environment-friendly, clean energy resource. In the near future, the biomass energy is likely to be one of the most dominant REs, contributing to a substantial reduction in consumption of carbon-emitting fossil fuels and electricity for cooking as well as lighting. Many methods have been adopted to produce combined heat energy and power (CHP), in improved cookstoves with advanced features of the flat thermoelectric generator (FTEG). Easy availability, accessibility, low-cost per capita produced energy, reduced emissions and inbuilt driving force have led to greater attention to CHP cookstove systems. Subsequently, exploring an efficient biomass advanced cookstove technology is necessary. An accurate inbuilt electricity generation unit to drive forced draft combustion fan/blower, to optimise performance with high efficiencies, is proposed in this current study.

For the previous two decades, the use of thermoelectric generators and its applications have been investigated and improved worldwide due to its significant advantages of straight conversion of thermal energy into electricity with minimum/no moving parts and reduced noise. Hence, newly developed thermoelectric generators (TEGs) are accepted as green sustainable technology and are widely used as flexible generators for a diversified applications [1, 2]. A combined advanced micro-gasifier cookstove with the TEG for heat and power cogeneration from a single system is presented here.

Champier et al. [3] have developed a prototype of a biomass cookstove with a higher-efficient combustion chamber (CC), and TEG modules are attached to either side of the cookstove. It was found from the experimental investigations that 6 W of electrical power was generated from four numbers of TEG. Experimental analysis was performed by Nuwayhid et al. [4] on the biomass domestic wood stove coupled with natural convection-cooled thermoelectric generator. It was obtained from the results that about 4.2 W of power was produced from each component. Kraemer et al. [5] built an innovative solar flat thermal panel for electricity generation adopting Seebeck effect concept with higher thermal concentration. There was an interesting result obtained from the study that an electrical convention efficiency of 4.6% for 1000 W/m^2 solar conditions is about eight times much more powerful than the other systems. Omer and Infield [6] proposed a conceptual model of a TEG for the estimation of an optimal device in power production. By means of using their developed model, four different TEGs were compared. Besides, they have developed a two-stage solar concentrator for a cogeneration system (combined heat and power from the thermoelectric modules). It was proven from the studies conducted by Omer and Infield [7] that an improvement of thermal efficiency as well as overall characteristics of the solar thermal concentrator for combined heat and thermoelectric power generation can be achieved using second stage concentrator.

Atik [8] studies the thermoelectric performances of the concentrating collector, receiver and TEG modules. The electric power generation efficiency, system efficiency and surface area temperature of the receiver system were obtained from the different solar radiation (W/m^2) and from different concentration ratios. Chen [9] suggested a model to examine the conceptual efficiency of solar thermoelectric generators (STEGs), which includes thermal concentration as well as optical concentration. It was obtained from his study that the component efficiency increases with increase in hot side temperature, but thermal efficiency decreases with increasing hot side temperature. Furthermore, he stated that the STEG efficiency can be improved when it has been maintained under evacuated condition.

Manikandan and Kaushik [1] performed the energy and exergy study of the solar annular thermoelectric generator (SATEG) in view of the impact of Thomson effect in concurrence with Peltier effect, Joule effect and Fourier heat conduction. Their study proposed annular thermoelectric generator as an alternative for flat thermoelectric generator (FTEG) to increase the cross-sectional area along the radial direction. They have established that the power output (W) and whole exergy efficiency (%) of the SATEG were 1.92 W and 5.02%, respectively, which was established to be 0.52 and 0.40% greater than that of SFTEG. They suggested that this SATEG system could be effectively used as the thermal insulation material as it offered better heat transfer characteristics; it was simple to drive and maintain compared to the solar flat plate thermoelectric generator.

From an in-depth literature survey, it is established that there are certainly no research studies available on CHP system in cookstove and an ATEG. Studies have explored only the performance evaluation of energy efficiency and exergy efficiency of the cookstove with an FTEG. The advantage of using an ATEG instead of FTEG is that it enhances the heat transfer surface area on both hot side and cold side, due to cylindrical structure. It is proposed to study the behaviour of ATEG placed outside the cylindrical CC to convert waste heat (hot side) into electricity as well as dissipate heat (cold side) by passing and preheating air through annular fins. Hence, it is indeed essential to study the energy and exergy analysis of an advanced cookstove-based ATEG for clean combustion and self-sustained cooking options, particularly suited for developing countries. Here, the exergy analysis (second law of thermodynamics) is an advanced method to enumerate the real irreversibilities delivered in the thermodynamic process.

2. Materials and methods

2.1 System description

The design and fabricated CC of the proposed biomass advanced micro-gasifier cookstove [10–15] are clearly depicted in **Figure 1**. The CC is fabricated using carbon steel with an inner diameter of 110 mm and an inner height of 155 mm. By means of increase in the proportion percentage (%) of carbon content, the cylinder material becomes harder and stronger providing higher creep properties. Carbon steel with high carbon content has been used to ensure it can withstand high temperatures.

The secondary air (combustion air) injection of the proposed cookstove is skewed to an angle of 45° towards bottom grate [11, 12, 16]. This is to confirm the better turbulence in the course of volatile combustion and for the period of char combustion. Exfoliated vermiculite mineral matter of 93% with glass wool of 2%

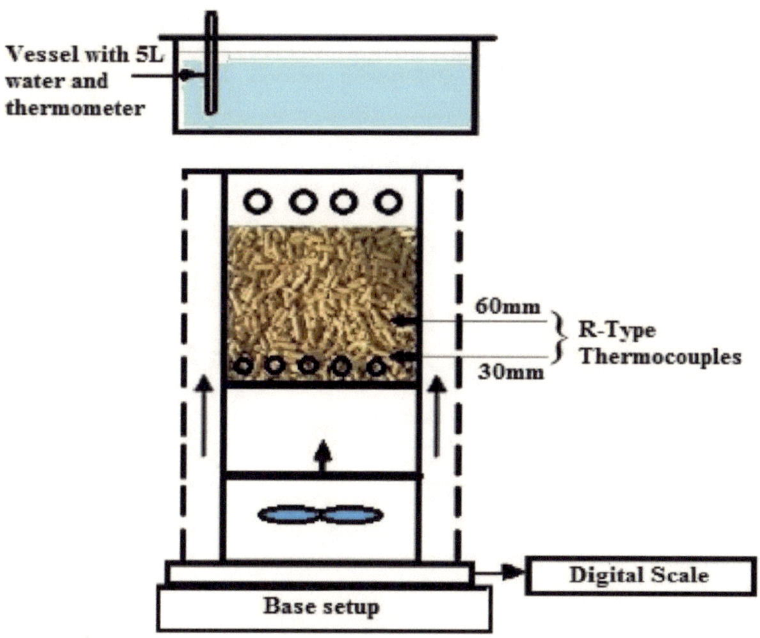

Figure 1.
Model of the proposed cookstove.

and Portland cement of 5% altogether by weight is moulded into a mixture for the preparation of the thermal protection lining material inside the concentric cylindrical CC. The air gap between two concentric cylindrical rings of the CC is filled with the same refractory composite paste. The thermal conductivity (k) of the prepared composite blend slab is calculated to be about 0.047 W/m°C by performing steady-state thermal conductivity test, as suggested by BIS-IS 9489 [17]. Due care is taken to ensure the total absence of any bypass channels due to the faulty lining of the thermal composite insulation material in the gasification and combustion air paths. A blower of the capacity 12 V DC is fitted below the combustion chamber of the cookstove. It tends to force the ambient air upwards along the way through the side of the CC; gasification and combustion ducts of diameter 4 and 3 mm provide the needed gasification and combustion air.

In the advanced micro-gasifier cookstove-based ATEG system, the exterior surface of the CC is in connection with the hot side junction of the GATEG. Hence, the waste heat ejected or lost is effectively used by the ATEG for the generation of electric power. The remaining heat energy available at the cold side junction of ATEG is exploited for preheating the secondary combustion air. The primary heat generated by the advanced micro-gasifier system is used for cooking food on the stove. This combined cogeneration cookstove system can deliver both electric power to drive fan/blower (also lighting, micro-charger applications) and cooking applications in rural areas from biomass energy. The graphical illustration of the combined biomass advanced micro-gasifier cookstove with ATEG system is shown in **Figure 2(a** and **b)**.

2.2 Characterisation of fuels

The two different categories of biomass used in the study are:

a. Solid biomass

(i) *Prosopis juliflora (Seemai Karuvelam)*

(ii) Coconut shells *(Thotti/Cherattai/Kottanguchi)*

b. Pellet (densified biowaste)

(iii) Tamarind seed pellet *(Puliyan kottai)*

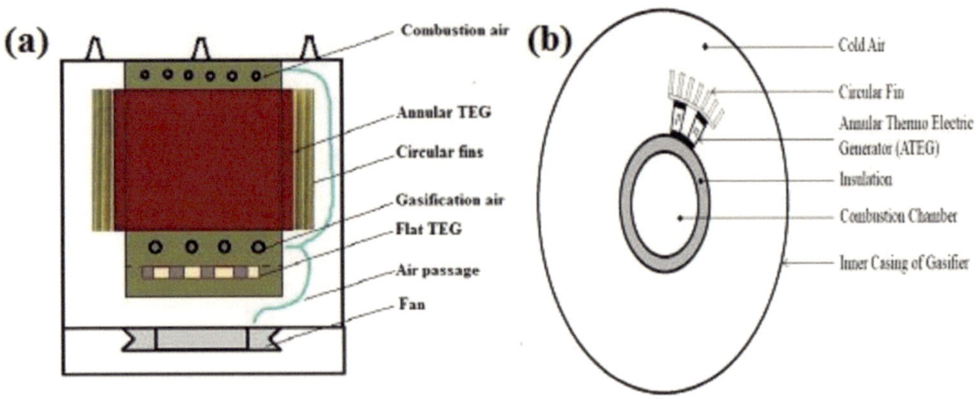

Figure 2.
(a) Advanced micro-gasifier annular thermoelectric generator system and (b) cross-sectional view.

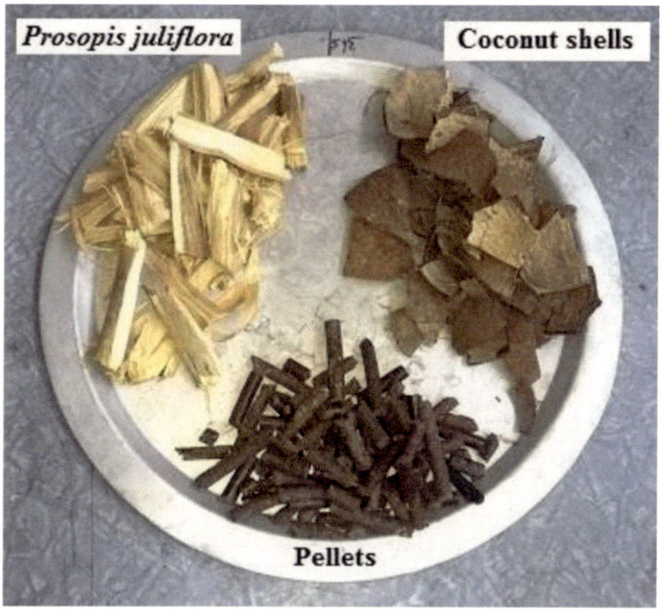

Figure 3.
Photograph of the biomass fuels considered for the experimental study.

Characteristics	*Prosopis juliflora*	Coconut shells	Tamarind seed pellet	Standard
Size (cm^3)	5 × 2 × 0.5	5 × 3 × 0.25	5 × 1	–
Bulk density (kg m^{-3})	560±20	610±20	1200±10	–
HHV (MJ/kg)	17.7	17.37	16.2	ASTM E711-87
Moisture content (%)	5.4	10	10.07	ASTM E871-82
Volatile matter (%)	77.9	72.05	63.02	ASTM E872-82
Ash content (%)	1.0	0.59	9.07	ASTM D1102-84
Fixed carbon (%)	15.7	17.34	18.04	*By difference*
Carbon (%)	45.5	45.84	50.15	ASTM E777-08
Hydrogen (%)	6.4	5.51	6.02	ASTM E777-08
Nitrogen (%)	0.6	0.35	0.42	ASTM E778-08
Oxygen (%)	47.2	47.58	41.41	*By difference*
Sulphur (%)	0.3	–	0.28	–

Table 1.
Proximate and ultimate analysis of the three different fuels [11–15]

The three types of biomass with different densities and ash content considered for this work are depicted in **Figure 3**, including two types of solid biomass wood with different properties and an agro residue seed pellet. All the solid biomass and pellets used are first dried in sunlight for 24 hours to ensure uniform moisture content (5–10%).

Table 1 summarises the properties of all the types of biomass used. Pictures of all the biomass types used in the experiments are illustrated in **Figure 3**.

3. Thermodynamic modelling

3.1 Thermodynamic modelling of the advanced micro-gasifier cookstove

The graphical illustration of the advanced micro-gasifier cookstove with the ATEG is presented in **Figure 4**. The thermal resistance network of the combined advanced cookstove ATEG is exposed in **Figure 5**.

By using the first law of thermodynamics, the energy balance equation of the biomass advanced cookstove ATEG can be written as

$$Q_{gen} = Q_{loss} + Q_{useful} \tag{1}$$

$$Q_{gen} = m_{fuel} \times CV_{fuel} \tag{2}$$

$$Q_{loss} = Q_{rad} + Q_{cond} + Q_{conv} \tag{3}$$

$$Q_{rad} = \sigma \varepsilon_{pan} A_{pan} \left({T_{pan}}^4 - {T_{air}}^4 \right) \tag{4}$$

$$Q_{cond} = Q = \frac{T_{cc} - T_h}{R_{s1} + R_{ins} + R_{s2} + R_{teg}} \tag{5}$$

$$Q_{conv} = \frac{T_h - T_c}{R_{air}} \tag{6}$$

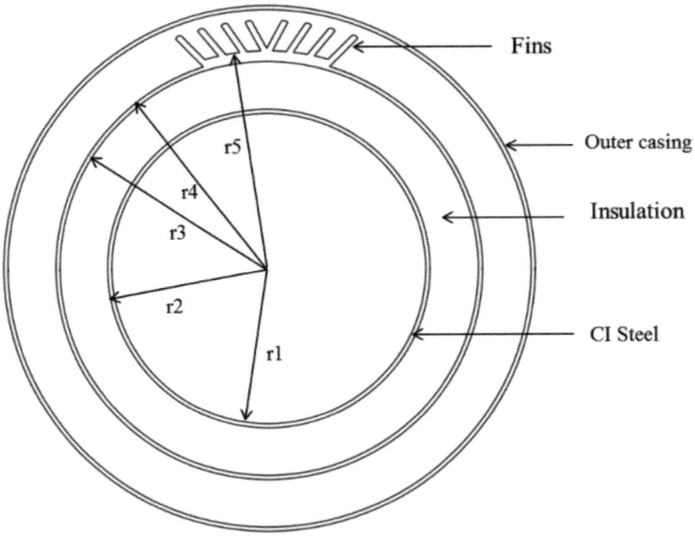

Figure 4.
Schematic illustration of the advanced micro-gasifier ATEG.

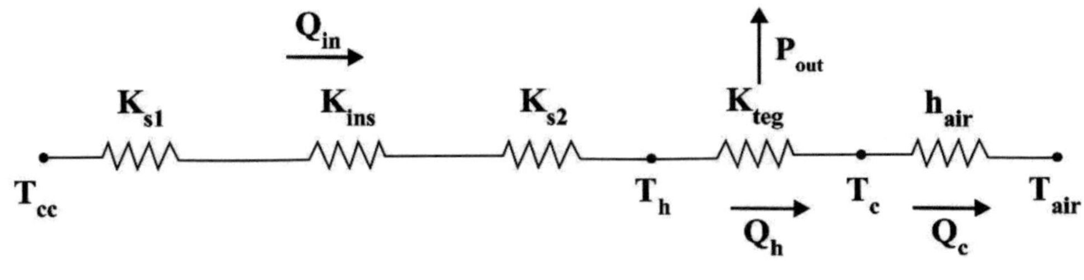

Figure 5.
The thermal network of advanced micro-gasifier ATEG.

3.1.1 Conductive resistance

Conductive resistance from a CC to ATEG (radial direction) is determined considering one-dimensional (1D) with steady-state conductive heat transfer via composite cylindrical walls. Three different types of materials, namely, steel-1 and steel-2 made up of extraordinary carbon steel, are used; the thermal insulation material used is Vermiculite composite:

$$R_{s1} = \frac{1}{2\pi k_{s1} L} ln \frac{r_2}{r_1} \tag{7}$$

$$R_{ins} = \frac{1}{2\pi k_{ins} L} ln \frac{r_3}{r_2} \tag{8}$$

$$R_{s2} = \frac{1}{2\pi k_{s2} L} ln \frac{r_4}{r_3} \tag{9}$$

$$R_{teg} = \frac{1}{2\pi k_{teg} L} ln \frac{r_5}{r_4} \tag{10}$$

3.1.2 Convective resistance

Convective resistance from ATEG to combustion air (radial direction) is considered as 1D steady-state convective heat transfer for ATEG (through extended surface finned annulus):

$$R_{air} = \frac{1}{2\pi h_{air} L r_5} \tag{11}$$

3.1.3 Heat transferred at hot and cold side junction of ATEG

Waste heat from the outer surface of CC is absorbed by the ATEG (Q_h) at the hot junction can be deliberated by the energy balance equation given as

$$Q_h = \alpha \mathrm{I} T_h - \frac{I^2 R}{2} + K(T_h - T_c) - \frac{\tau I (T_h - T_c)}{2} \times n \tag{12}$$

The transmitted heat through the ATEG is released at the cold side junction (annular fins) of the ATEG by preheating the combustion air. The equation is written as

$$Q_c = \alpha \mathrm{I} T_c + \frac{I^2 R}{2} + K(T_h - T_c) + \frac{\tau I (T_h - T_c)}{2} \times n \tag{13}$$

3.2 Thermodynamic modelling of ATEG

A cross-sectional observation of the thermoelement of an ATEG is depicted in **Figure 6**. The cross-sectional area A(r) of the ATEG thermoelectric pair increases in a radial direction (r). The exhaustive thermodynamic modelling and energy and exergy analysis of the ATEG in view of the Thomson effect have been deliberated in the following section, as studied by Kaushik and Manikandan [18]. The assumptions used for the thermodynamic modelling and investigations of ATEG are:

- 1D steady-state heat transfer equation of ATEG alongside the radial path is deliberated for the study.

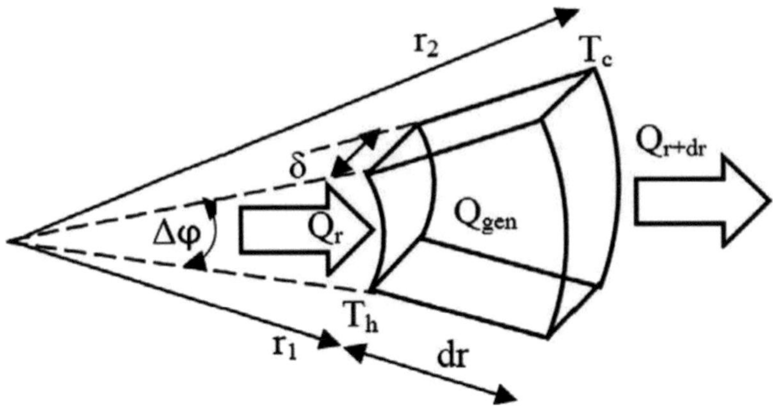

Figure 6.
Cross-sectional view of ATEG [1].

- The thickness (δ) of the ATEG module is constant throughout.
- Convection losses and radiation losses from the sides of the thermoelectric modules to the atmosphere are negligible (as heated air is recirculated into the combustion chamber).
- The electrical resistance of the contact is presumed to be about 10% of the actual inbuilt electrical resistance.

For the study, it has been assumed that $Q_{storage} = Q_{g, loss} = 0$; however Q_r and Q_{r+dr} are the heat input (from waste heat) supplied to the ATEG from outside the CC and heat output dissipated from the ATEG into the secondary air, respectively, whereas Q_{gen} can be the addition of Thomson and Joule's heat produced in the element (dr) [1]. The cross-sectional region of the thermoelement is established on the study conducted by Shen et al. [19].

Based on the assumptions, the cross-sectional area, length (L) and thickness (δ) of the p-type and also n-type thermoelectric (TE) leg are the same; the dispersal of temperature in the p-type and n-type leg of the ATEG is also assumed to be the same. Shen et al. [19] have studied the thermal conductance (K) and electrical resistance (R) of the ATEG are as given below:

$$K = (K_n + K_p) = \frac{\Delta\varphi\delta}{ln(r_5/r_4)}(k_n + k_p) \tag{14}$$

$$R = (R_n + R_p) = \frac{ln(r_5/r_4)}{\Delta\varphi\delta}\left(\rho_n + \rho_p\right) \tag{15}$$

The only difference is the value of K and R in the thermodynamic modelling of the FTEG and ATEG; the rest of the equations for the GATEG and GFTEG are comparable with Shen et al. (2015):

$$I = \frac{(\propto - \tau)(T_h - T_C)}{R + R_L} \tag{16}$$

$$R_L = \sqrt{1 + ZT_m R} \tag{17}$$

The thermal properties and electrical properties of a TE material combined together are referred to as figure of merit (FOM). Dimensionless FOM has been generally used to measure the desirability of TE materials for devices by multiplying with mean operating temperature (T_m) [1]:

$$ZT_m = \frac{\propto(\propto - \tau)}{\rho k}\frac{(T_h - T_C)}{2} \tag{18}$$

The power output (W) produced, electrical energy efficiency (%) and exergy efficiency (%) of an advanced micro-gasifier-based ATEG system can be considered from the altered work done by Manikandan and Kaushik [1]:

$$P_{out} = Q_h - Q_c = (\alpha - \tau)(T_h - T_c)I - I^2R = I^2R_L \tag{19}$$

Eq. (19) designates that the Thomson effect will decrease the power output of the ATEG. The energy efficiency (electrical) of advanced cookstove-assisted ATEG is given as

$$\eta_{el} = \frac{P_{out}}{Q} = \frac{(\alpha - \tau)(T_h - T_c)I - I^2R}{Q_{cond}} \times n \tag{20}$$

The exergy efficiency (electrical) of the advanced cookstove-assisted ATEG is derived as

$$\psi_{el} = \frac{P_{out}}{E_Q} = \frac{(\alpha - \tau)(T_h - T_c)I - I^2R}{Q_{cond}\left(1 - \frac{T_{air}}{T_h}\right)} \times n \tag{21}$$

Hence, the combination of potential energy and exergy efficiencies of the advanced micro-gasifier cookstove ATEG system can be written as

$$\text{Combined efficiency (\%)} = \eta_{el} + \eta_{th} \tag{22}$$

The energy as well as exergy analysis of the micro-GATEG is analysed via engineering equation solver (EES) for different operating conditions.

3.3 Cookstove performance

3.3.1 Energy efficiency

Thermal efficiency (%) is defined as the fraction of heat energy given off by the biomass fuel that is successfully transported to the water in the cooking vessel. The remaining unrecovered heat energy is dissipated into the largest heat sink of an atmosphere. The method used to assess the thermal energy efficiency is specified in Eq. 23, as follows:

$$\eta th = \left\{ \frac{\left[4.186 \times (m_{wi} - m_{wf}) \times (T_{wf} - T_{wi})\right] + (W_v \times 2257)}{m_{fw} \times CV_{fuel}} \right\} \tag{23}$$

3.3.2 Exergy efficiency

The maximum possible work, which can be created by a system for a particular environmental condition, is generally taken as the Carnot hypothetical maximum relating to the ambient temperature. The thermal exergy input supplied to the pot for water boiling can be stated as below [13, 14, 20]:

$$\text{Exin} = (m\text{TP} \times C1 \times \eta c + \text{x} \times d \times \text{C2})\left(1 - \frac{Ta}{T\text{fuel}}\right) \tag{24}$$

The exergy output of the ATEG attached advanced micro-gasifier cookstove is the quantity of energy spent by the boiling water times the Carnot factor as follows [20]:

$$\text{Ex}o = \left\{mw \times Cp \times (T\text{fw} - \text{Tiw})\left(1 - \frac{Ta}{T\text{fw}}\right) + \text{mpot} \times Cp, \text{pot} \times (T\text{fp} - \text{Tip})\left(1 - \frac{Ta}{T\text{fp}}\right)\right\} \tag{25}$$

Generally, lowering heat source or raising heat sink lowers exergy. The exergy efficiency (ψ) is well defined as the fraction between the output exergy and the input exergy as shown below:

$$\eta\text{th} = \frac{\text{Ex}o}{\text{Exin}} \tag{26}$$

A small number of essential parameters like the mass of water (kg), the weight of fuel (kg), the volume of the kerosene sample (for the ignition of fuel) and the weight of unfilled Al vessel were computed before starting the test. Readings of water temperature (°C) and pot temperature (°C) were taken on a minute-to-minute basis. The reference conditions taken for exergy analysis are To=303 K and Po=101.325 kPa.

4. Results and discussions

4.1 Analysis of energy and exergy efficiencies of the advanced micro-gasifier cookstove

The performance of the ACS cookstove is assessed in terms of energy efficiency (%) and exergy efficiency (%) using *Prosopis juliflora*, coconut shells and tamarind seed pellets as fuel. It is perceived that the thermal efficiencies of the stove are 36.7±0.4, 37±0.4 and 35±0.4% for coconut shell, *Prosopis juliflora* and tamarind pellets, respectively, after four repetition tests. The exergy efficiencies (%) of the cookstove are 15.6±0.45, 17.5±0.45 and 15±0.45% for the discussed three different fuels. The uncertainties for energy and exergy efficiencies are established as 0.43% and 0.48%, respectively.

The comparison on the energy efficiency and exergy efficiency of the advanced micro-gasifier cookstove is illustrated in **Figure 7** for a distinct set of operational constraints. It is also witnessed that the energy efficiency (%) performance of the ACS cookstove is considerably higher than that of exergy efficiency (%) performance. This is due to the extent of energy extracted in the hot water for ACS cookstove being much less than the worth of energy extracted due to temperature constraint; this phenomenon is common for all cookstoves.

4.2 Conceptual modelling results of GATEG

In this investigative study, the energy and exergy analysis of an advanced micro-gasifier cookstove ATEG is studied under various operating circumstances. The

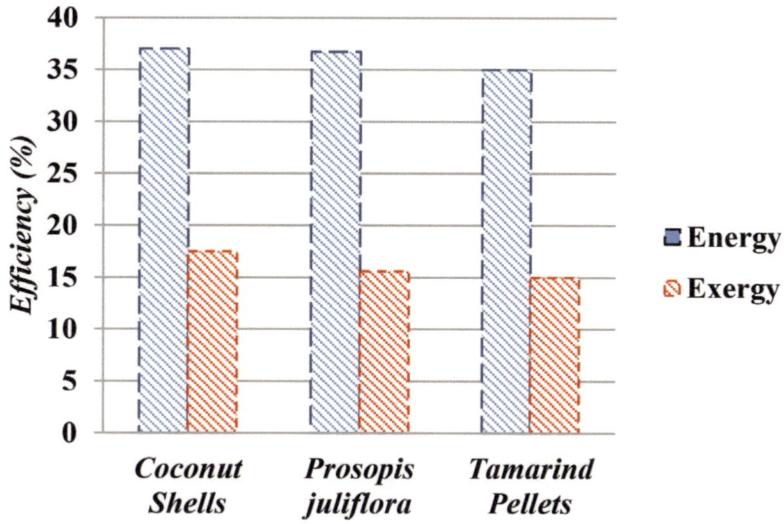

Figure 7.
Energy and exergy efficiencies of the advanced micro-gasifier cookstove.

influence of hot side and cold side junction temperature and the influence of the number of thermocouples and operating electric current (A) on electric power output (W), thermal output (of cookstove) and energy and exergy efficiencies (%) of the GATEG system are studied.

4.2.1 Effect of change in hot junction temperature

The influence of CC temperature outside the insulation on the power output as well as the energy and exergy efficiencies (%) of the ATEG have been studied. The combustion air temperature passed over the annular fins of the TEG varies with time so the cold junction temperature is also varied between 30 and 150°C in this study. The atmospheric temperature is considered as 30°C.

The effects of various hot junction temperatures on the power output of the ATEG are shown in **Figure 8(a–f)**. The various zones of the CC temperature determine the hot side temperature of the ATEG. During combustion of the advanced micro-gasifier cookstove, the flame front propagates downwards with respect to the fuel bed density and heat transfer rate. Hence, heat is not uniform throughout from top to bottom. The uniform temperature is reached only after the conversion takes place from volatile combustion mode to char combustion mode, almost at the end (after 70% of weight loss in fuel). Therefore, when there is a variation in hot junction temperature, like an increase from a smaller value to higher, the power output (W) and the optimum current (A) value for maximum power output also increase. The reason is that when the temperature inside the combustion zone increases, a respective temperature of the outside chamber after insulation increases, thus increasing the power output of the TEG. It is also clear from **Figure 8(a–f)** that the power output of the micro-gasifier annular thermoelectric generator is maximum when the outside combustion temperature is at a maximum of 275°C. At a current flow rate of 0.8 A, the actual power output of GATEG is 10.05 W, electrical energy efficiency is 6.76% and exergy efficiency is 15%.

Similarly, from **Figure 8(a–f)**, it is clear that the power output of the micro-gasifier ATEG is minimum when the outside combustion temperature is at a minimum of 150°C. At a current flow rate of 0.4 A, the power output of GATEG is 2.414 W, electrical energy efficiency is 1.31% and exergy efficiency is 4.62%.

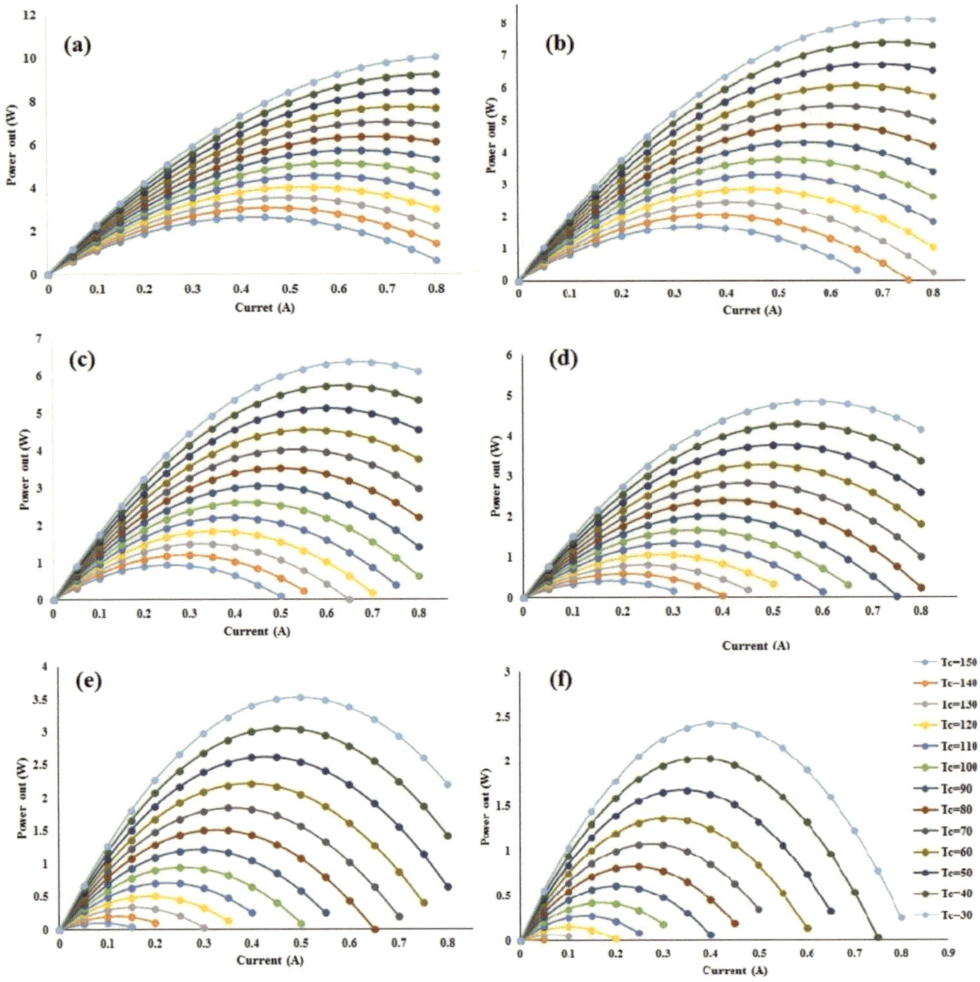

Figure 8.
Power output (W) with respect to current (A) for (a) 275°C, (b) 250°C, (c) 225°C, (d) 200°C, (e) 175°C and (f) 150°C.

The influence of variation of hot lateral temperature on the electrical energy efficiency (%) of GATEG is shown in **Figure 9(a–f)**. It can be observed from **Figure 9(a–f)** that the electrical energy efficiency of GATEG is for the range of heat input considered, for the maximum hot side temperature of 275°C and an operational current flow of 0.8 A; the electrical efficiency (%) of GATEG is 6.76%.

The variation of exergy efficiency of GATEG for changing cold junction temperatures at maximum hot side temperature of 275°C is shown in **Figure 10(a–f)**. It is seen that the exergy efficiency of the GATEG is high for all working conditions. It is obvious that the exergy efficiency of GATEG obtained for the heat input of 149 W at a working current of 0.8 A is 6.76%. This is due to the power output (i.e., exergy output) of the GATEG that is marginally greater because of superior heat transfer rates.

4.2.2 Effect of the number of thermocouples

The influence of the number of thermo-plates (i.e., thermocouples) on the performance variance like power output (W), energy efficiency and exergy

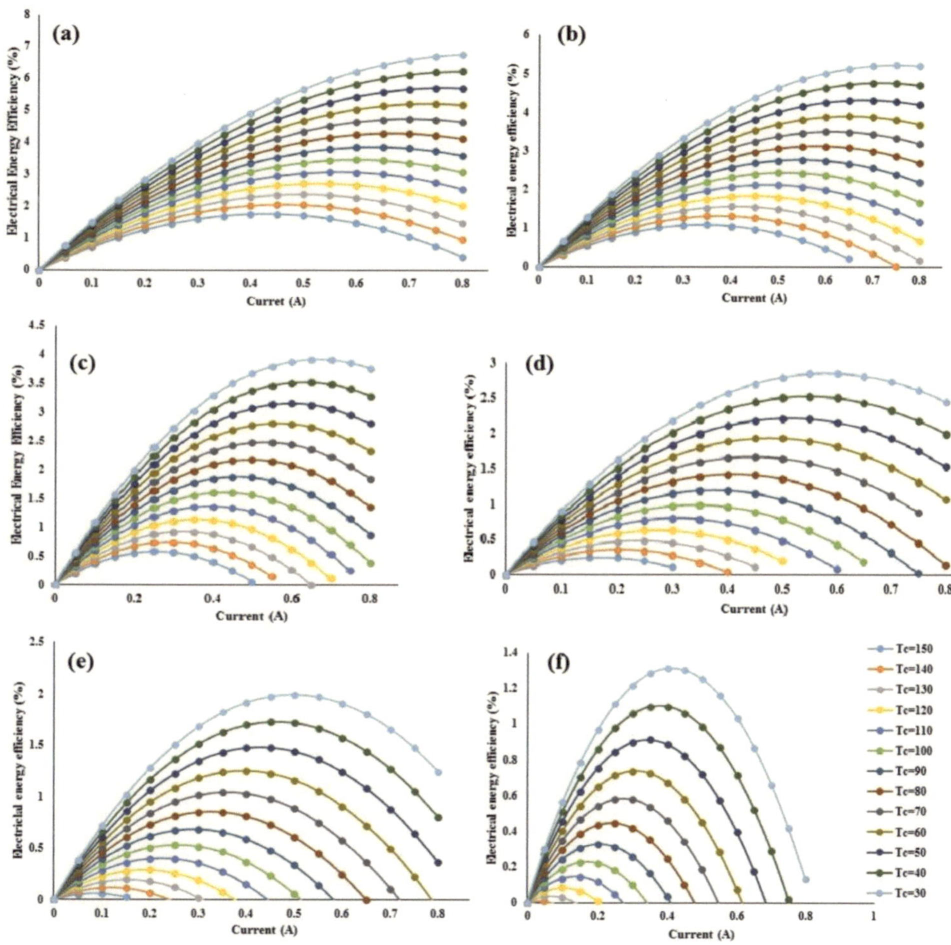

Figure 9.
Electrical energy efficiency (%) with respect to current (A) for (a) 275°C, (b) 250°C, (c) 225°C, (d) 200°C, (e) 175°C and (f) 150°C.

efficiency of the ACS cookstove annular/flat thermoelectric generator cogeneration system is studied. With an increase in the number of thermocouples, there is a rise in the heat transfer area. Hence, heat transfer between a hot side and cold side junction of the ATEG system is improved, as deliberated by Manikandan and Kaushik [1] and He et al. [21]. **Figure 11** shows the effect of numbers in thermocouples on the power output (W) of GATEG, with a clear indication of the number of thermocouples being directly proportional to the power output of GATEG, as proposed by Manikandan and Kaushik [1].

The hot junction temperature is considered as 275°C, and the cold side temperature is retained at 30°C. The losses in the systems are considered as negligible. **Figure 11** demonstrates the influence of the number of TEG modules on the electric power generation for variation of current (A) levels. **Figure 11** clearly indicates the power produced; it is maximum as a result of the addition of/rises in the number of TEG modules.

Eventually, the addition of modules leads to overall thermal resistance causing a reduction in the combined modules, R_{tem}, which leads to a fall in temperature difference, rapidly offsetting any further rise in voltage output. The plots shift to small current range slowly as "n" increases primly due to the interior electrical resistance upsurges steadily with the number of modules. The maximum power

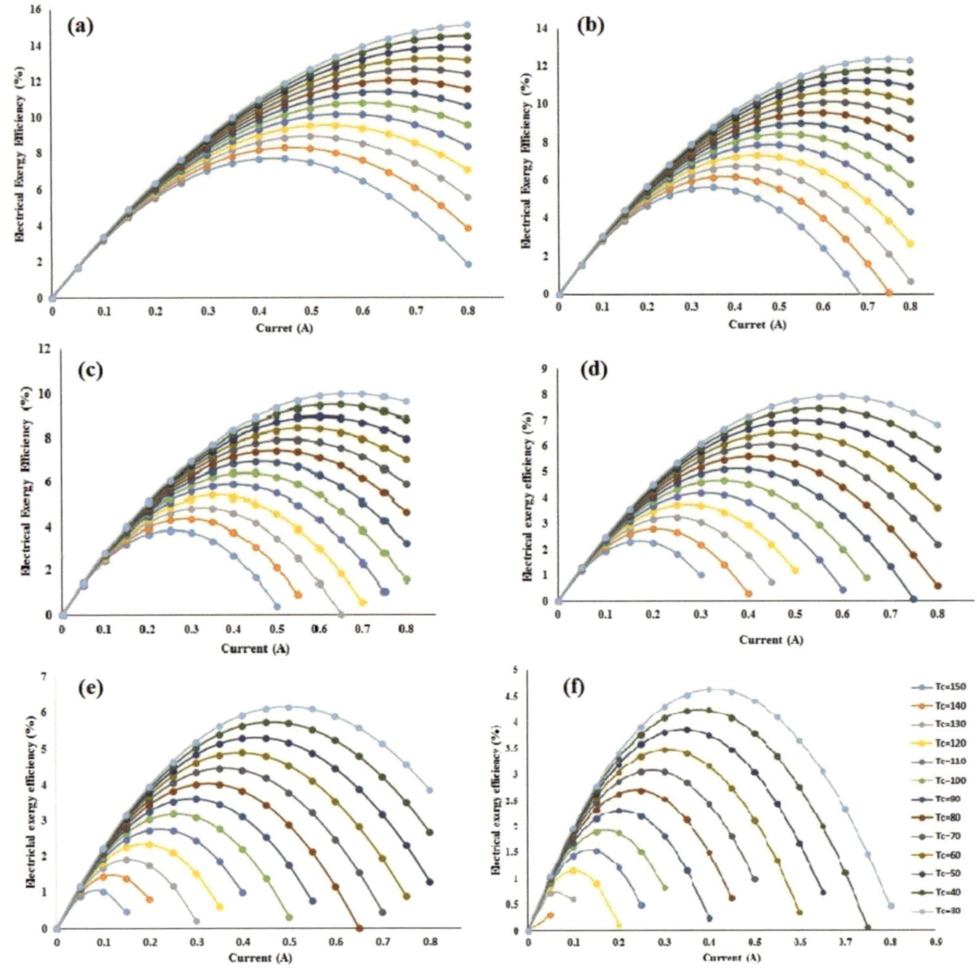

Figure 10.
Electrical exergy efficiency (%) with respect to current (A) for (a) 275°C, (b) 250°C, (c) 225°C, (d) 200°C, (e) 175°C and (f) 150°C.

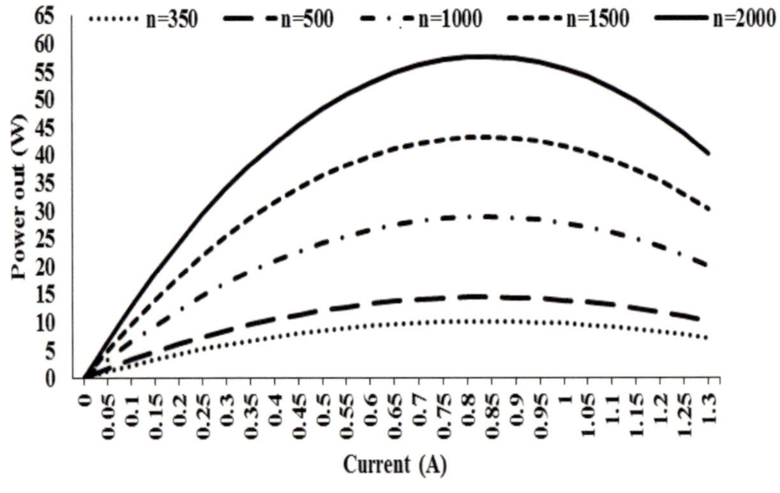

Figure 11.
Influence of the number of TEG modules on the power output.

(W) is obtained once the load resistance (R_L) matches with the system resistance according to maximum power transfer theorem. A similar curve with a notable difference in power output indicates that the maximum power output is attained for an increasing number of thermoelectric modules. The increase in efficiency by increasing 100 numbers into 1000 numbers is 100%. Thus, from the power output point of view, using an increased number of modules produces more power. This observation is similar to that of a steady state conducted by Jie Chen et al. [22]. A further intensification in the number of thermocouples results in an increase in surface area and volume which offers more resistance, thereby increasing the temperature of combustion air and consequently reducing the power output of TEG (refer **Figure 11**) and its electrical energy efficiency. These outcomes are comparable to those recorded by He et al. [21, 23].

5. Conclusions

An investigation of the micro-gasifier ATEG is conducted based on the first law and second law of thermodynamics, and its performance factors are investigated for varying hot and cold side temperature conditions and by varying number of thermocouples. From the conceptual modelling, the following conclusions are summarised:

- The power output (W), electrical energy efficiency (%), exergy efficiency (%) and combined system energy and exergy efficiencies (%) of the GATEG are 10.05 W, 6.76%, 15.12% and 43.46% and 30.72%, respectively, calculated for the maximum temperature difference of 275°C across the TEG, with the help of EES software.

- The same modelling is repeated for a low surface temperature of 150°C outside the combustion chamber. The power output (W), electrical energy efficiency (%), exergy efficiency (%) and combined system energy efficiency (%) of the GATEG are 2.41 W, 1.31%, 4.63% and 38% and 20.23%, respectively, for the maximum temperature difference of 150°C across the TEG.

- The advanced micro-gasifier cookstove annular thermoelectric generator is a suitable option since it has many advantages like enhanced heat transfer characteristics in the hot side and cold side of the ATEG due to the greater heat transfer area. The diameter (D) of the CC (based on the cooking load requirement) can be increased if necessary, to provide more heat transfer surface area to the ATEG.

The fixing of the ATEG with the cylinder-shaped CC will be very easy, and the facility to arrange for thermal insulation to the cold lateral of the TEG will become easier if the ATEG has been adopted. The conceptual model analysis untaken in this study may be supportive in designing actual GATEG systems for electric power production from engine exhaust heat (flue gas), other heat pipes, etc. As the power output (W) and overall exergy efficiency (%) of the GATEG are low, these arrangements are improvident, but with the improved/novel TEG materials with the higher ZT, this type of concepts will gain more significance in the near future.

Nomenclature

Q_{gen}	heat generated inside the combustion chamber of gasifier (W)
Q_{loss}	heat loss by conduction, convection and radiation (W)
Q_{useful}	the heat input to the purpose of water boiling (W)
m_{fuel}	mass rate of fuel consumed (kg/s)
CV_{fuel}	calorific value of wood (kJ/kg)
Q_{rad}	radiation loss offered by the pan (W)
σ	Stefan-Boltzmann constant (W/m^2K^4)
ε_{pan}	emissivity of the pan
A_{pan}	surface area of the pan (m^2)
T_{pan}	final temperature of the pan (K)
T_{air}	atmospheric air temperature (K)
$Q_{cond} = Q$	heat input to the ATEG (W)
T_{cc}	temperature inside the combustion chamber (K)
T_h	temperature at hot junction of ATEG (K)
R_{s1}	conductive resistance for steel-1 (K/W)
R_{ins}	conductive resistance for insulation (K/W)
R_{s2}	conductive resistance for steel-2 (K/W)
R_{teg}	conductive resistance for ATEG (K/W)
Q_{conv}	heat rejected by the cold junction of ATEG (W)
T_h	temperature at hot junction of ATEG (K)
T_c	temperature at cold junction of ATEG (K)
k_{s1}, k_{s2}	thermal conductivities of steel-1 and steel-2 (W/m K)
k_{ins}	thermal conductivity of insulation (W/m K)
K_{teg}	thermal conductivity of TEG (W/m K)
L	length of the combustion chamber or ATEG (m)
r_1	inner diameter of combustion chamber or steel-1 (m)
r_2	outer diameter of steel-1 and inner diameter of insulation (m)
r_3	outer diameter of insulation and inner diameter of steel-2 (m)
r_4	outer diameter of steel-2 and inner diameter of ATEG (m)
r_5	outer diameter of ATEG (m)
h_{air}	convective heat transfer coefficient (W/m^2K
Q_h	heat absorbed by the hot junction of the ATEG (W)
Q_c	heat rejected by the cold junction of the ATEG (W)
K	thermal conductance of the ATEG (W/K)
R	electrical conductance of the ATEG (Ω)
φ	angle (radians)
k_p, k_n	thermal conductivity of p and n legs of ATEG (W/m K)
ρ_n, ρ_p	electrical resistivity of p and n legs of ATEG (Ωm)
I	current (A)
α	Seebeck coefficient (V/K)
τ	Thompson coefficient (V/K)
R_L	external resistive load of imposed on ATEG (Ω)
ZT_m	dimensionless figure of merit
Z	figure of merit (1/K)
P_{out}	power output from the ATEG (W)
n	no of thermoelectric couples
η_{el}	electrical energy efficiency (%)
ψ_{el}	exergy efficiency (%)

Author details

Sakthivadivel Duraisamy[1*], Manikandan Sundararaj[3], Kumaraselvan Raja[2], Ganesh Kumar Poongavanam[2] and Iniyan Selvarasan[2]

1 School of Mechanical Engineering (SMEC), Vellore Institute of Technology (VIT) University, Vellore, Tamil Nadu, India

2 Department of Mechanical Engineering, Institute for Energy Studies, College of Engineering Guindy Campus, Anna University, Chennai, India

3 SRM Institute of Science and Technology, Tamil Nadu, India

*Address all correspondence to: sakthi2energy@gmail.com

References

[1] Manikandan S, Kaushik SC. Energy and exergy analysis of solar heat pipe based annular thermoelectric generator system. Solar Energy. 2016;**135**:569-577

[2] Karri MA, Thacher EF, Helenbrook BT. Exhaust energy conversion by thermoelectric generator: Two case studies. Energy Conversion and Management. 2011;**52**:1596-1611

[3] Champier D, Bedecarrats JP, Rivaletto M, Strub F. Thermoelectric power generation from biomass cookstoves. Energy. 2010;**35**:935-942

[4] Nuwayhid RY, Shihadeh A, Ghaddar N. Development and testing of a domestic woodstove thermoelectric generator with natural convection cooling. Energy Conversion and Management. 2005;**46**:1631-1643

[5] Kraemer D, Poudel B, Feng HP, Caylor JC, Yu B, Yan X, et al. High-performance flat-panel solar thermoelectric generators with high thermal concentration. Nature Materials. 2011;**10**:532-538

[6] Omer SA, Infield DG. Design optimization of thermoelectric devices for solar power generation. Solar Energy Materials & Solar Cells. 1998;**53**:67-82

[7] Omer SA, Infield DG. Design and thermal analysis of a two stage solar concentrator for combined heat and thermoelectric power generation. Energy Conversion and Management. 2000;**41**:737-756

[8] Atik K. Numerical simulation of a solar thermoelectric generator. Energy Source Part A. 2011;**33**:760-767

[9] Chen G. Theoretical efficiency of solar thermoelectric energy generators. Journal of Applied Physics. 2011;**109**: 104908-1041–8

[10] Raman P, Murali J, Sakthivadivel D, Vigneswaran VS. Performance evaluation of three types of forced draft cookstoves using fuel wood and coconut shell. Biomass and Bioenergy. 2013;**49**: 333-340

[11] Sakthivadivel D, Balakumar A, Iniyan S. Development of advanced cookstove with optimum air mixture using CFD. Asian Journal of Research in Social Sciences and Humanities. 2017; **7**(2):384-392

[12] Sakthivadivel D, Iniyan S. Combustion characteristics of biomass fuels in a fixed bed micro-gasifier cookstove. Journal of Mechanical Science and Technology. 2017;**31**(2): 995-1002

[13] Sakthivadivel D, Iniyan S. Characterization, density and size effects of fuels in an advanced micro-gasifier stove. Biofuels. 2018;**31**(2): 995-1002

[14] Sakthivadivel D, Iniyan S. Experimental design and 4E (energy, exergy, emission and economical) analysis of a fixed bed advanced micro-gasifier stove. Environmental Progress & Sustainable Energy. 2018;**37**(6): 2139-2147

[15] Sakthivadivel D, Ganesh Kumar P, Iniyan S. Computational modeling and performance evaluation of an advanced micro-gasifier cookstove with optimum air injection. Biofuels. 2019. https://doi.org/10.1080/17597269.2019.1573606

[16] Tryner J, Tillotson JW, Baumgardner ME, Mohr JT, MW DF, Marchese AJ. The effects of air flow rates, secondary air inlet geometry, fuel type, and operating mode on the performance of gasifier cookstoves. Environmental Science & Technology. 2016;**50**:9754-9763

[17] BIS. Method of test for thermal conductivity of thermal insulation materials by means of heat flow meter (First Revision of IS 9489). In: Bureau of Indian Standards. 2015

[18] Manikandan S, Kaushik SC. Energy and exergy analysis of an annular thermoelectric cooler. Energy Conversion and Management. 2015;**106**: 804-814

[19] Shen ZG, Wu SY, Xiao L. Theoretical analysis on the performance of annular thermoelectric couple. Energy Conversion and Management. 2015;**89**:244-250

[20] Tyagi SK, Pandey AK, Sahu S, Bajala V, Rajput JPS. Experimental study and performance evaluation of various cookstove models based on energy and exergy analysis. Journal of Thermal Analysis and Calorimetry. 2013;**111**: 1791-1799

[21] He W, Su Y, Wang Y, Riffat S, Ji J. A study on incorporation of thermoelectric modules with evacuated-tube heat-pipe solar collectors. Renewable Energy. 2012;**37**(1):142-149

[22] Chen J, Zuo L, Wu Y, Klein J. Modeling, experiments and optimization of an on-pipe thermoelectric generator. Energy Conversion and Management. 2016;**122**: 298-309

[23] He W, Su Y, Riffat S, Hou J, Ji J. Parametrical analysis of the design and performance of a solar heat pipe thermoelectric generator unit. Applied Energy. 2011;**88**(12):5083-5089